Robert G

Differential Geometry

1972 Lecture Notes

MINKOWSKI
Institute Press

Robert Geroch
Enrico Fermi Institute
University of Chicago

Cover: Lecture notes are often written in similar environments

ISBN: 978-1-927763-06-3 (softcover)
ISBN: 978-1-927763-07-0 (ebook)

Minkowski Institute Press
Montreal, Quebec, Canada
http://minkowskiinstitute.org/mip/

For information on all Minkowski Institute Press publications visit our website at http://minkowskiinstitute.org/mip/books/

Contents

1 Manifolds 1

2 Smooth Functions: Smooth Mappings 5

3 Vectors at a point 9

4 Linear Algebra 13

5 Tensors at a point 19

6 Tensor Fields 21

7 Lie Derivatives 25

8 Symmetrization and Antisymmetrization 33

9 Exterior Derivatives 35

10 Derivative Operators 41

11 Concomitants 45

12 Curvature 49

13 Metrics 53

14 Curvature Defined by a Metric 59

15 Smooth Mappings: Action on Tensor Fields 63

16 Bundles 71

17 The Tensor Bundles 75

18 Smooth Mappings: Action on Tensor Bundles 79

19 Curves 81

20 Integral Curves 83

21 The Lie derivative: Geometrical Interpretation 89

22 Lie Groups 93

23 Groups of Motions 97

24 Dragging Along 99

25 Derivative Operators: Interpretation in the Tensor Bundles 103

26 Riemann Tensor: Geometrical Interpretation 107

27 Geodesics 109

28 Submanifolds 111

29 Tangents and Normals to Submanifolds 113

30 Metric Submanifolds 115

31 Tensor Fields and Derivatives on Submanifolds 119

32 Extrinsic Curvature 123

33 The Gauss-Codazzi Equations 127

34 Metric Submanifolds of Co-Dimension One 131

35 Orientation 135

36 Integrals 137

37 Stokes' Theorem 141

38 Integrals: The Metric Case 145

Appendix 149

About the Author 153

1. Manifolds

The arena on which all the action in differential geometry takes place is a mathematical object called a manifold. In this section, we shall define manifolds and discuss a few of their properties. A good intuitive understanding of what a manifold is, as well as what can and cannot be done on a manifold, is essential for differential geometry.

Roughly speaking, a manifold is a space having the "local smoothness structure" of R^n. [R^n, or Euclidean $n-$space, is the set consisting of $n-$tuples, $(x_1, x_2, \ldots, x_n)$, of real numbers.] The idea, then, is to isolate, from all the rich structure on R^n, (e.g., its metric structure, its vector-space structure, its topological structure), that one small bit of structure we call "smoothness." [This process of isolating separate structures for individual attention is, of course, common in mathematics.]

Let M be a set. By an *n−chart* on M we understand a subset U of M and a mapping $\varphi : U \to R^n$ having the following two properties: i) φ in one-to-one [That is, no two distinct points of U are taken to the same point of R^n by φ.], and ii) The image of U, that is, the subset $0 = \varphi[U]$ of R^n, is open in R^n. [Recall that a subset 0 of R^n is said to be open if, for any $x \geq 0$, there is a number $\epsilon > 0$ such that, whenever $d(x, y) < \epsilon$, y is also in 0, where $d(x, y)$ is the usual Euclidean distance in R^n.]

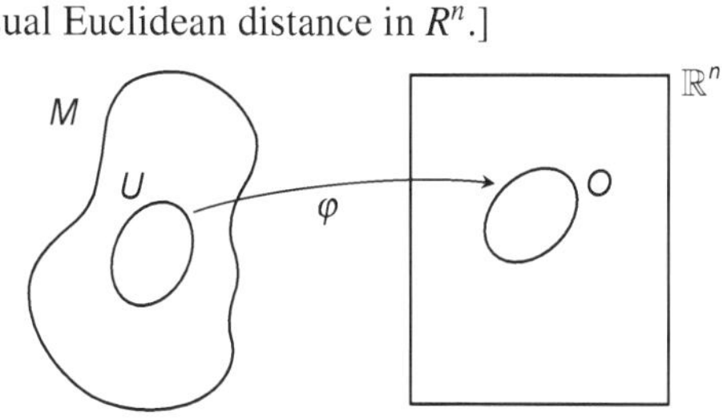

These charts are the mechanism by which we intend to induce our "local smoothness structure" on the set M. They are well-suited for the job. A chart, after all, sets up a correspondence between a set of points U of M and an (open) set of points 0 of R^n. Thus, a chart defines n real-valued functions on U. The points of U can be labeled by the values of these n functions. Since no two distinct points of U have all the same values, such a labeling is unambiguous. (Charts are sometimes called "coordinate patches.") To obtain a manifold, we must lay down enough charts on M, and require that, when two charts overlap (i.e., when their U's overlap in M), the corresponding smoothness structures agree.

Let (U, φ) and (U', φ') be two n–charts on the set M. If U and U' intersect in M, then there is induced on their intersection, $U \cap U'$, two smoothness structures. One is obtained from φ [which defines a one-to-one correspondence between $U \cap U'$ and the subset $\varphi[U \cap U']$ of R^n], and the other from φ' [which defines a one-to-one correspondence between $U \cap U'$ and the subset $\varphi'[U \cap U']$ of R^n. By composing these, we obtain a correspondence, defined by $\varphi' \cdot \varphi^{-1}$ and its inverse, $\varphi \cdot \varphi'^{-1}$, between $\varphi[U \cap U']$ and $\varphi'[U \cap U']$ – both subsets of R^n. It is this correspondence which is used to compare the "smoothness structures." The n–charts (U, φ) and (U', φ') on M are said to be *compatible* if the following hold: i) $\varphi[U \cap U']$ and $\varphi'[U \cap U']$ are open subsets of R^n. ii) the mappings $\varphi' \cdot \varphi^{-1} : \varphi[U \cap U'] \to \varphi'[U \cap U']$ and $\varphi \cdot \varphi'^{-1} : \varphi'[U \cap U'] \to \varphi[U \cap U']$ are C^∞. [A mapping from a subset of R^n to R^n is just another way of speaking of n functions of n variables. Such a mapping is said to be C^∞ if all the partial derivatives of all n functions exist and are continuous.]

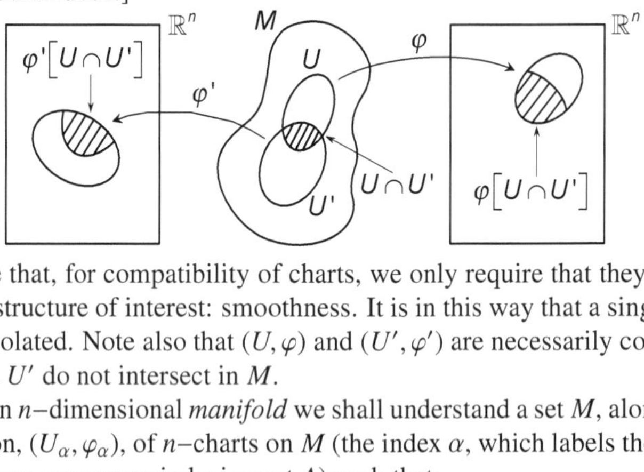

Note that, for compatibility of charts, we only require that they agree in the one structure of interest: smoothness. It is in this way that a single structure is isolated. Note also that (U, φ) and (U', φ') are necessarily compatible if U and U' do not intersect in M.

By an n–dimensional *manifold* we shall understand a set M, along with a collection, $(U_\alpha, \varphi_\alpha)$, of n–charts on M (the index α, which labels the various charts, runs over some indexing set A) such that:

M1. Any two charts in the collection are compatible.

M2. The charts cover M, i.e., $\bigcup_\alpha U_\alpha = M$.

M3. Any n–chart, which is compatible with all the charts in the collection, is itself in the collection.

M4. If p and p' are distinct points of M, then there exist charts $(U_\alpha, \varphi_\alpha)$ and (U_β, φ_β) such that p is in U_α, p' is in U_β, and such that U_α and U_β do not intersect.

These conditions – or at least the first three of them – are exactly the conditions one would expect. Intuitively, the condition $M1$ states that "if two charts induce a smoothness structure in the same region of M, then those structures must agree." The second condition states that "every part of M has induced on it a smoothness structure." The third condition ensures that we have not placed too much structure on M by cutting down the number of charts. (For example, R^n, which has a much richer structure than that of a manifold, is defined by just "one chart.") Since compatibility looks only at the smoothness structure, and since, by $M3$, we admit all the compatible charts, we conveniently "wash out" all the other structure. The rather ugly fourth condition eliminates certain pathological objects (called non-Hausdorff manifolds) which are not very interesting.

Thus, a manifold consists of a set M along with some charts on that set. We shall usually refer to a manifold as "M," the charts understood. The following sentence is intended as a guide to the literature: What we have called an "n–dimensional manifold" might be referred to by others as a "C^∞, Hausdorff, n–dimensional manifold, without boundary, which is not necessarily paracompact or connected."

One quickly builds up a sufficiently strong intuition about manifolds that it becomes unnecessary to constantly refer back to conditions $M1 - M4$. The discussion which follows is intended to help in this process.

It might seem that, because of condition $M3$, it would be almost impossible to specify a manifold, for the collection of all those compatible charts will be an extremely large one. The possibility of describing manifolds easily arises from the following fact:

Remark 1. Let M be a set on which there is specified a collection of n–charts satisfying $M1$, $M2$, and $M4$. Let C denote the collection of of all n–charts which are compatible with those of the given collection. Then the set M, with the charts C, defines a manifold.

Exercise 1: Prove Remark 1. (Everything is clear except condition $M1$, which is not difficult to establish.) Intuitively, Remark 1 states that it is easy to reduce the structure. If you have a set covered by a compatible collection of charts, you can always reduce the structure down to that of a manifold by including more charts.

We give two examples of manifolds.

Example 1. Let M consists of n–tuples of real numbers, so, as a point-set, $M = R^n$. We introduce some charts on M: let U be any subset of M which, considered as a subset of R^n, is open, and let φ be the identity mapping from U to R^n. These are certainly charts, and they clearly satisfy $M1$, $M2$, and $M4$. By Remark 1, we obtain a manifold based on M. This manifold is called the manifold R^n.

Example 2. Let M consist of $(n+1)$–tuples of real numbers, $(y_1, \ldots, y_n)$, which satisfy $(y_1)^2 + (y_2)^2 + \ldots + (y_{n+1})^2 = 1$. We introduce some charts on M. Let U_{1^+} consist of points of M with $y_1 > 0$, and let φ_{1^+}, acting on the point $(y_1, \ldots, y_{n+1})$ of U_{1^+} be the point $(y_2, \ldots, y_{n+1})$ of R^n. Let U_{1^-} consist of the points of M with $y_1 < 0$, and let φ_{1^-}, acting on $(y_1, \ldots, y_{n+1})$ be the point

$(y_2, \ldots, y_{n+1})$ of R^n. Let U_{2^+} consist of points of M with $y_2 > 0$, and let φ_{2^+}, acting on the point $(y_1, \ldots, y_{n+1})$ of U_{2^+}, be the point $(y_1, y_3, y_4, \ldots, y_{n+1})$. Continuing in this way, we obtain $(2n + 2)$ charts, (U_{i^+}, φ_{i^+}) and (U_{i^-}, φ_{i^-}), $i = 1, 2, \ldots, (n + 1)$. These n–charters are all compatible with each other, and cover M. By including all charts compatible with these, we obtain an n–dimensional manifold called the n–sphere, S^n.

Exercise 2. Verify that Example 2 leads to a manifold.

There are a number of techniques for obtaining new manifolds from old ones. We describe two such techniques.

Let M and M' be manifolds, of dimensions n and n', respectively. We define a new manifold, $\widetilde{M}$, which is called the *product* of M and M', written $\widetilde{M} = M \times M'$. The dimension of $\widetilde{M}$ will be $(n + n')$. As a point-set, $\widetilde{M}$ is the product of the sets M and M'. That is to say, a point of $\widetilde{M}$ consists of a pair, (p, p'), with p in M and p' in M'. We must introduce some charts on this set M. Let $(U_\alpha, \varphi_\alpha)$ be a chart on M, and $(U_{\alpha'}, \varphi_{\alpha'})$ a chart on M'. Then, since U_α is a subset of M and $U_{\alpha'}$ a subset of M', $\widetilde{U} = U_\alpha \times U_{\alpha'}$ is a subset of $\widetilde{M}$. Now φ_α maps U_α to R^n, while $\varphi_{\alpha'}$ maps $U_{\alpha'}$ to $R^{n'}$. Now let p be a point of U_α, and p' a point of $U_{\alpha'}$, so the pair (p, p') is a point of $\widetilde{U}$. Then $\varphi_\alpha(p)$ is some point, say $(x_1, \ldots, x_n)$, of R^n, while $\varphi_\alpha(p')$ is some point, $(y_1, \ldots, y_{n'})$, of $R^{n'}$. Thus, we can associate with the point (p, p') of $\widetilde{U}$ the point $(x_1, \ldots, x_n, y_1, \ldots, y_{n'})$ of $R^{n+n'}$. In other words, we have just described a mapping, φ, from $\widetilde{U}$ to $R^{n+n'}$. This is a chart on $\widetilde{M}$. Thus, given a chart on M and a chart on M', we obtain a chart on $\widetilde{M}$. It is easy to check that, since M and M' are manifolds, these charts on $\widetilde{M}$ satisfy conditions $M1$, $M2$, and $M4$. (Exercise 3. Do so.) By Remark 1, we have a manifold M.

Example 3. The manifold $R^1 \times S^1$ is called the cylinder. The manifold $S^1 \times S^1$ is called the torus. The manifold $R^n \times R^{n'}$ is the same as the manifold $R^{n+n'}$.

A second method of constructing manifolds is by "cutting holes" in known manifolds. However, only certain types of "holes" can be cut without destroying the manifold structure. Let M be a manifold, and let S be a subset of M. S is said to be *open* in M if, for any point p of S, there is a chart (U, φ), such that p is in U and U is a subset of S. A set is said to be *closed* if its complement is open. (Needless to say, these are the open sets for a topology on M. This topology could also be defined, for example, as the coarsest for which the chart maps are continuous.) Now let C be a closed subset of the manifold M. We define a new manifold $\widetilde{M}$, written $M - C$, which might be called "M with C cut out". As a point-set, $\widetilde{M}$ consists of those points of M which do not lie in C. As charts on $\widetilde{M}$, we take those charts $(U_\alpha, \varphi_\alpha)$ on M for which U_α does not intersect C. Exercise 4. Verify that this M is a manifold, i.e., check conditions $M1 - M4$.

A reasonable fraction of the manifolds which occur in physics can be obtained from the manifold R^n and S^n using the operations of taking products and cutting holes.

2. Smooth Functions: Smooth Mappings

Let M be a manifold, and let f be a (real-valued) function on M. Then, if (U, φ) is a chart on M, f is, in particular, a function on U. Hence, $f \cdot \varphi^{-1}$ is a function on the subset $\varphi[U]$ of R^n. In other words, $f \cdot \varphi^{-1}$ is just a real-valued function of n variables. The function f on M is said to be *smooth* if $f \cdot \varphi^{-1}$ is C^∞ for every chart (U, φ). [Note this technique. To define something on a manifold, one uses the charts to express that something in terms of R^n, where we already know what it means.]

Exercise 5. Let f be a function on M for which $f \cdot \varphi^{-1}$ is C^∞ for some collection of charts satisfying $M2$. Verify that f is a smooth function on M.

Exercise 6. Verify that the function x_1 is smooth on R^n; that the function y_1 is smooth on S^n.

Exercise 7. Let $F(z_1, \ldots, z_k)$ be a C^∞ real-valued function of k real variables. Let $f_1, \ldots, f_k$ be k smooth functions on a manifold M. Show that $F(f_1, \ldots, f_k)$ is a smooth function on a M.

Exercise 8. Let f be a smooth function on M, f' a smooth function on M'. Define a function $\tilde{f}$ on $\widetilde{M} = M \times M'$ as follows: if p is a point of M, p' a point of M', $\tilde{f}(p, p') = f(p) + f'(p')$. Prove that this $\tilde{f}$ is a smooth function on $\widetilde{M}$.

Exercise 9. Let f be a smooth function on M and C a closed subset of M. Prove that $\tilde{f}$, restricted to $\widetilde{M} = M - C$, is a smooth function on $\widetilde{M}$.

We denote by $\mathfrak{F}$ the collection of smooth functions on M.

There is a sense in which the smooth functions on a manifold M characterize the manifold structure of M. This is made more precise by the following:

Remark 2. Let M be a set, and let C and C' be manifold structures on M. (That is, each of C and C' is a collection of charts on M satisfying $M1 - M4$.) Suppose that every smooth functions on (M, C) is also a smooth function on (M, C'). Then $C = C'$.

Exercise 10. Prove Remark 2. (It is perhaps not surprising that one can give axioms on a collection of functions on a set which are necessary and sufficient for those functions to be the smooth functions for some manifold structure on the set.)

Things in mathematics often come in pairs. One introduces some class of objects of interest (e.g., groups, topological spaces, vector spaces, or sets). Then, one introduces certain natural, structure-preserving mappings (sometimes called morphisms) between objects in the same class (e.g., homomorphisms of groups, continuous mappings on topological spaces, linear mappings on vector spaces, or just plain mappings on sets). In the same spirit, manifolds are objects. The corresponding morphisms, which we now define are called differentiable, or smooth mappings.

It is convenient to define a smooth mapping using the smooth functions. [Since the smooth functions "characterize" the manifold structure, it is not surprising that a reasonable definition can be expressed in this way.] Let M and M' be manifolds, and let $\psi : M \to M'$ be a mapping from the point-set M to the point-set M'. If f' is any smooth function on M', then $f' \cdot \psi$ is a function on M. [In words, $f' \cdot \psi$ consists of the following instructions: given a point p of M, take p to M' using ψ, and evaluate f' at the corresponding point, $\psi(p)$ of M'.] The mapping $\psi : M \to M'$ is called a *smooth mapping* if, for every smooth function f' on M', $f' \cdot \psi$ is a smooth function on M.

Exercise 11. Let M be a manifold, and let f be a (real-valued) function on M. Then this f certainly defines a mapping from the manifold M to the manifold R^1. Prove that the function is smooth if and only if the mapping of manifolds is smooth. [Thus, smooth functions can be considered as a special case of smooth mappings.]

Exercise 12. Reexpress the definition of a smooth mapping directly in terms of charts.

Exercise 13. Let $\widetilde{M} = M \times M'$. Define a mapping $\psi : \widetilde{M} \to M$ as follows. If p is a point of M, and p' is a point of M', so (p, p') is a point of $\widetilde{M}$, then $\psi(p, p') = p$. Prove that ψ is smooth.

Exercise 14. Let $\widetilde{M} = M \times M'$. Define a mapping $\psi : M \to \widetilde{M}$ as follows. Fix a point p' of M'. Then, if p is any point of M, $\psi(p) = (p, p')$. Prove that ψ is smooth.

Exercise 15. Let C be a closed subset of the manifold M. Define a mapping ψ from $M - C$ to M as follows. If p is a point of $M - C$, $\psi(p) = p$. Prove that ψ is smooth.

Morphisms always have the property that the composition of two is another. Smooth mappings are no exception.

Theorem 3. Let M, M', and M'' be manifolds, and let $\psi : M \to M'$ and $\psi' : M' \to M''$ be smooth mappings. Then $\psi' \cdot \psi : M \to M''$ is a smooth mapping.

Proof: Let f'' be a smooth function on M''. We must show that $f'' \cdot \psi' \cdot \psi$, is a smooth function on M. Since f'' is smooth on M'' and ψ' is a smooth mapping, $f'' \cdot \psi'$ is a smooth function on M'. But, since $f'' \cdot \psi'$ is a smooth function on M' and ψ is a smooth mapping, $f'' \cdot \psi' \cdot \psi$ is a smooth function on M.

Exercise 16. What are the compositions of the mappings in Exercises 13 and 14?

An important concept in mathematics is that of an isomorphism between objects. Roughly speaking, two objects are isomorphic if they are identical as far as the structure under consideration is concerned. [For groups and vector spaces, isomorphisms are called isomorphisms; for topological spaces, homeomorphisms; for sets, bijections.] The "isomorphisms" for manifolds are called differeomorphisms. Let M and M' be manifolds, and let $\psi : M \to M'$ be a smooth mapping. This ψ is said to be a *diffeomorphism* if ψ is one-to-one and onto (so that ψ^{-1} is well-defined), and $\psi^{-1} : M' \to M$ is also a smooth mapping. Given two manifolds, M and M', there may or may not exist a diffeomorphism between them. If there does exist one, M and M' are said to be *diffeomorphic*. One occasionally writes $M = M'$ to indicate that M and M' are diffeomorphic.

Example 4. Define a mapping $\psi : S^n \to R^n$ as follows. If $(y_1, \ldots, y_{n+1})$ is any point of S^n, $\psi(y_1, \ldots, y_{n+1})$ is the point $(y_1, \ldots, y_n)$ of R^n. Although this ψ is smooth, it is not one-to-one: for example, $\psi(0, \ldots, 0, -1) = \psi(0, \ldots, 0, 1)$. Hence, ψ is not a diffeomorphism.

Example 5. Let $\psi : R^1 \to R^1$ be the smooth mapping $\psi(x) = x^3$. Since this ψ is one-to-one and onto its inverse, $\psi^{-1}(x) = x^{1/3}$, exists. But, since the inverse is not smooth, ψ is not a diffeomorphism.

Example 6. Let C be the closed subset of R^2 consisting of the single point $(0, 0)$. Define a mapping ψ from $R^2 - C$ to $R^2 - C$ as follows: $\psi(x_1, x_2) = (r^{-2}x_1, r^{-2}x_2)$, where $r^2 = (x_1)^2 + (x_2)^2$. This ψ is clearly a diffeomorphism.

Exercise 17. Verify that $S^n - C$, where C is the point $(0, \ldots, 0, 1)$ of S^n, is diffeomorphic with R^n. Verify that $R^n - C$, where C is the point $(0, \ldots, 0)$ of R^n, is diffeomorphic with $S^{n-1} \times R^1$.

Exercise 18. Prove that the composition of two diffeomorphisms is a diffeomorphism. Prove that the inverse of a diffeomorphism is a diffeomorphism.

Exercise 19. Prove that, if two manifolds are diffeomorphic, then they have the same dimension.

As a final example to illustrate the structure possessed by a manifold, we establish a theorem to the effect that manifolds are "locally homogeneous." [The theorem is somewhat analogous to the following fact about vector spaces: If v and w are non zero elements of a vector space V, then there is an isomorphism from V to V which takes v to w.] Thus, manifolds, like vector spaces, are barren, devoid of landmarks.

Theorem 4. Let p be a point of the manifold M. Then there is an open set O, containing p, with the following property: given any point p' in O, there exists a diffeomorphism $\psi : M \to M$ with $\psi(p) = p'$. Proof: Let (U, φ) be a chart such that p is in U. Let $\varphi(p) = z = (z_1, \cdots, z_n)$, a point of R^n.

Choose $\epsilon > 0$ such that, whenever x is a point of R^n with $d(x, z) < \epsilon$, x is in $\varphi[U]$. Denote by V the collection of all x in R^n with $d(x, z) < \epsilon$. The subset $O = \varphi^{-1}[V]$ of M is our candidate for the O in the theorem. [Exercise 20. Why is O open in M?]

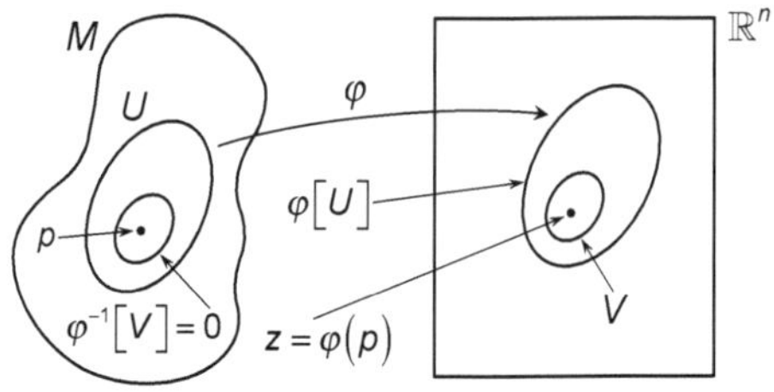

Let p' be a point of O, so $\varphi(p') = z' = (z'_1, \cdots, z'_n)$, a point of R^n, is in V. Set $\epsilon' = d(z, z') < \epsilon$. Define a mapping Λ from V to V as follows:

$$\Lambda(x_1, \cdots, x_n) = (x_1, \cdots, x_n) + f(r)(z'_1 - z_1, \cdots, z'_n - z_n),$$

where we have set $r = [d(x, z)]$, and where $f(r)$ is a function of one variable with the following properties: i) $f(r)$ is C^∞, ii) there is an $\epsilon_1 > 0$ such that $f(r) = 1$ for $r < \epsilon_1$, iii) there is an $\epsilon_2 < \epsilon$ such that $f(r) = 0$ for $r > \epsilon_2$, and iv) $|df/dr| < (\epsilon')^{-1}$. [Since $\epsilon < \epsilon$, there exists such an $f(r)$.] These conditions ensure that Λ is a smooth mapping from V to V (conditions i and ii), that $\Lambda(z) = z'$ (condition ii), that Λ is the identity near the edge of V (condition iii), and that Λ^{-1} exists and is smooth (conditions i, ii, iii, and iv). [Exercise 21. Verify these properties.]

We now define the mapping ψ from M to M. Let q be a point of M. If q is not in O, set $\psi(q) = q$. If q is in O, set $\psi(q) = \varphi^{-1} \cdot \Lambda \cdot \varphi(q)$. It is clear that this ψ is smooth, one-to-one, onto, and that $\psi(p) = p'$. [Exercise 22. Check that ψ is a diffeomorphism.]

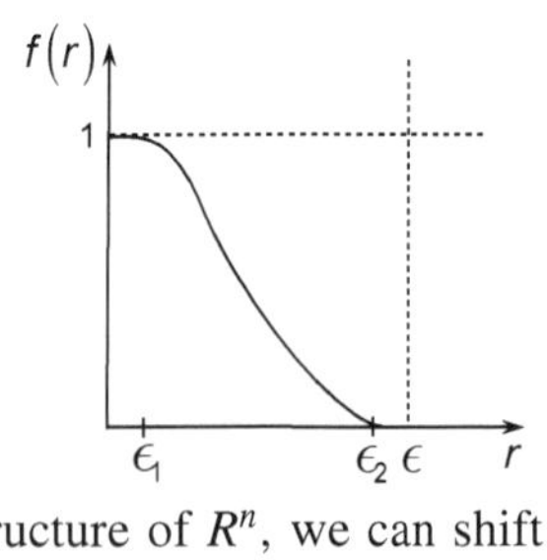

Intuitively, since we can shift points a little bit in R^n, preserving the smoothness structure (a fact made explicit by the expression, above, for $\Lambda(x)$), and since M has the local smoothness structure of R^n, we can shift points a little in M, preserving the smoothness structure.

The statement of Theorem 4 is interesting because of what it says about manifolds. The proof of Theorem 4 is interesting because it illustrates a useful technique: certain local properties of R^n can, using the charts, be pulled back to properties of manifolds.

Exercise 23. State and prove: S^n and R^n are globally homogeneous.

3. Vectors at a point

We now have the arena for differential geometry: a manifold. The natural things which exist in such an environment are objects called tensor fields. These tensor fields will be defined in the next few sections. Although the fields are by far the most important, it is convenient to first introduce tensors at a point.

We begin with vectors at a point, $\underline{x} = (\underline{x}^1, \ldots, \underline{x}^n)$ of R^n. A vector in R^n at the point $\underline{x}$ can be represented by its components, $(\xi^1, \ldots, \xi^n)$, with respect to the axes in R^n. That is to say, a vector in R^n at $\underline{x}$ is equivalent to n real numbers. However, while this representation of vectors is natural in R^n, it is not a very convenient one for discussing vectors on manifolds (in which there are no natural "axes"). Denote by $\mathfrak{F}(R^n)$ the collection of all smooth functions on R^n so an element of $\mathfrak{F}(R^n)$ consists of a C^∞, real-valued function, $f(x^1, \ldots, x^n)$, of n real variables. If $(\xi^1, \ldots, \xi^n)$ is the components of any vector at $\underline{x}$, and f is any smooth function on R^n, we set

$$\xi(f) = \xi^1 \left.\frac{\partial f}{\partial x^1}\right|_{\underline{x}} + \cdots + \xi^n \left.\frac{\partial f}{\partial x^n}\right|_{\underline{x}} \tag{1}$$

This $\xi(f)$, which is just a real number, is called the directional derivative of f in the direction of the vector represented by $(\xi^1, \ldots, \xi^n)$. It follows immediately from the elementary properties of the derivative that $\xi(f)$ satisfies the following conditions:

DD1. $\xi(f + g) = \xi(f) + \xi(g)$.
DD2. $\xi(fg) = f(\underline{x})\xi(g) + g(\underline{x})\xi(f)$.
DD3. If f = const., then $\xi(f) = 0$.

Thus, every vector in R^n at $\underline{x}$ defines, via (1), a mapping $\xi(f)$ from $\mathfrak{F}(R^n)$ to R, satisfying DD1, DD2, and DD3. We have :

Proposition 5. Eqn. (1) defines a one-to-one correspondence between n–tuples, $(\xi^1, \ldots, \xi^n)$ and mappings from $\mathfrak{F}(R^n)$ to R satisfying DD1, DD2, and DD3.

Proof: Two things must be shown: i) if $(\xi^1, \ldots, \xi^n)$ and $(\eta^1, \ldots, \eta^n)$ have the property that $\xi(f) = \eta(f)$ for all f in $\mathfrak{F}(R^n)$ (with $\xi(f)$ and $\eta(f)$ defined by Eqn. (1)), then $\xi^1 = \eta^1, \ldots, \xi^n = \eta^n$, and ii) if $\mu(f)$ is a mapping from $\mathfrak{F}(R^n)$

to R, satisfying DD1, DD2, and DD3, then $\mu(f)$ is of the form (1) for some $(\mu^1, \ldots, \mu^n)$.

i) Let $\xi(f) = \eta(f)$ for all f. Settings $f = x^1$, we have, from Eqn. (1), $\xi' = \xi(x^1) = \eta(x^1) = \eta'$. Similarly for $x^2, \ldots, x^n$.

ii) Let $\mu(f)$ satisfy DD1, DD2, and DD3. Define n numbers, $\mu^1 = \mu(x^1), \ldots, \mu^n = \mu(x^n)$. We shall show that, with this $(\mu^1, \ldots, \mu^n)$, Eqn. (1) holds for all f. Let f be in $\mathfrak{F}(R^n)$, and write f in the form

$$f(x) = f(\underline{x}) + (x^1 - \underline{x}^1)g_1(x) + \cdots + (x^n - \underline{x}^n)g_n(x) \tag{2}$$

with $g_1(x), \ldots, g_n(x)$ in $\mathfrak{F}(R^n)$. [See Lemma 6 below.] Then, from DD1 and DD2,

$$\begin{aligned} \mu(f) = \mu(f(\underline{x})) &+ (x^1 - \underline{x}^1)|_{\underline{x}}\, \mu(g_1) + g_1(x)|_{\underline{x}}\, \mu(x^1 - \underline{x}^1) \\ &+ \cdots + (x^n - \underline{x}^n)|_{\underline{x}}\, \mu(g_n) + g_n(x)|_{\underline{x}}\, \mu(x^n - \underline{x}^n) \end{aligned} \tag{3}$$

By DD3, $\mu(f(\underline{x})) = 0$; by DD1 and DD3, $\mu(x^i - \underline{x}^i) = \mu^i$; by Eqn.(2), $g_i|_{\underline{x}} = \partial f/\partial x^i|_{\underline{x}}$; clearly, $(x^i - \underline{x}^i)|_{\underline{x}} = 0$ $(i = 1, 2, \ldots, n)$. Eqn. (1) now follows from Eqn. (3).

In the proof of Proposition 5, we made use of the following fact:

Lemma 6. If f is in $\mathfrak{F}(R^n)$, then f can be written in the form (2) (where $\underline{x}$ is some fixed point of R^n) for some $g_1(x), \ldots, g_n(x)$ in $\mathfrak{F}(R^n)$.

Proof:

$$\begin{aligned} f(x) - f(\underline{x}) &= \int_0^1 \frac{\partial}{\partial t} f(\underline{x} + t(x - \underline{x}))dt \\ &= (x^1 - \underline{x}^1) \int_0^1 \frac{\partial}{\partial x^1} f(\underline{x} + t(x - \underline{x}))dt \\ &\quad + \cdots + (x^n - \underline{x}^n) \int_0^1 \frac{\partial}{\partial x^n} f(\underline{x} + t(x - \underline{x}))dt \end{aligned}$$

Proposition 5 provides a different, and more fruitful, representation of vector in R^n.

With the remarks above as motivation, we now return to manifolds. Let M be a manifold, and, as in Sect. 2, let $\mathfrak{F}$ be the collection of smooth functions on M. Then (e.g., by Exercise 7), the (pointwise) sum and product of elements of $\mathfrak{F}$ is in $\mathfrak{F}$. [This $\mathfrak{F}$ has the structure of a ring.] Fix a point p of M. By a *contravariant vector* in M at p, we mean a mapping $\xi : \mathfrak{F} \to R$ satisfying DD1, DD2, and DD3 (with, of course, the $\underline{x}$ in DD2 replaced by p). The collection of all contravariant vectors in M at p will be written $\mathfrak{F}^{\cdot}(p)$. If ξ and η are in $\mathfrak{F}^{\cdot}(p)$, m is a real number, and f is in $\mathfrak{F}$, set $(\xi + m\eta)(f) = \xi(f) + m\eta(f)$. With this as the definition of the sum of two elements of $\mathfrak{F}^{\cdot}(p)$, and of the product of an element of $\mathfrak{F}^{\cdot}(p)$ with a real number, $\mathfrak{F}^{\cdot}(p)$ clearly becomes a vector space.

The contravariant vectors can be represented more explicitly in terms of charts. Let (U, φ) be a chart on M. Then, if f is a smooth function on M, $f \cdot \varphi^{-1} = \tilde{f}(x^1, \ldots, x^n)$ is a smooth function on $\varphi[U]$. Now suppose that the point p is in U, and set $\underline{x} = \varphi(p)$. Choose $h^1, \ldots, h^n$ in $\mathfrak{F}$ such that, in some open subset of R^n containing $\underline{x}, \tilde{h}^1 = (x^1 - \underline{x}^1), \ldots, \tilde{h}^n = (x^n - \underline{x}^n)$. [Exercise 24. Why do such h's exist?] If ξ is a contravariant vector in M at p, the numbers $\xi' = \xi(h^1), \ldots, \xi^n = \xi(h^n)$ are called *components* of ξ with respect to the chart (U, φ).

Theorem 7. The components of ξ are independent of the choice of the h's. Furthermore, given n numbers, $(\xi', \ldots, \xi^n)$, there exists precisely one contravariant vector in M at p with components $(\xi', \ldots, \xi^n)$,

Proof: Let f be in $\mathfrak{F}$, and set

$$f = f(p) + g_1 h^1 + \cdots + g_n h^n + s^2 t \tag{4}$$

where $g_1, \ldots, g_n, s, t$ are in $\mathfrak{F}$ and $s(p) = 0$. [To construct such an expression, proceed as follows. First choose $g_1, \ldots, g_n$ so $f = f(p) + g_1 h^1 + \ldots + g_n h^n$ in some open set containing p. Then choose any s which is positive except at p, where it vanishes. Finally, choose t so that (4) is satisfied.] Using the argument surrounding Eqn. (3), and the fact that $\xi(s^2 t) = 2s(p)t(p)\xi(s) + [s(p)]^2 \xi(t) = 0$, we obtain

$$\xi(f) = \xi^1 \left.\frac{\partial \tilde{f}}{\partial x^1}\right|_{\underline{x}} + \cdots + \xi^n \left.\frac{\partial \tilde{f}}{\partial x^n}\right|_{\underline{x}} \tag{5}$$

That the components are independent of the h's is clear from Eqn. (5). To prove the second statement, observe that, if $(\xi^1, \ldots, \xi^n)$ are n numbers, then Eqn. (5) defines a contravariant vector in M at p with components $(\xi^1, \ldots, \xi^n)$. Thus, if M is an n-dimensional manifold, and p is a point of M, then $\mathfrak{F} \cdot (p)$ is an n-dimensional vector space.

The components assigned to a vector depend, of course, on the choice of chart. We derive the well-known formula for this dependence. Let (U, φ) and (U', φ') be two charts, both containing the point p of M. Then $U\hat{U}'$ is coordinated in two ways, once by φ and once by φ' . That is to say, we have a smooth, one-to-one, onto mapping, $\varphi' \cdot \varphi^{-1}$, from $\varphi[U\hat{U}']$ to $\varphi'[U\hat{U}']$, both subsets of R^n. In other words, we have a functions of n variables, $x'^1(x^1, \ldots, x^n), \ldots, x'^n(x^1, \ldots, x^n)$. Let f be a smooth function on M. Then, setting $\tilde{f} = f \cdot \varphi^{-1}$, $\tilde{f}' = f \cdot \varphi'^{-1}$, we have $f(x^1, \ldots, x^n) = f'(x'^1(x^1, \ldots, x^n), \ldots, x'^n(x^1, \ldots, x^n))$. Now let ξ be a contravariant vector in M at p and let $(\xi^1, \ldots, \xi^n)$ and $(\xi'^1, \ldots, \xi'^n)$ be the components of ξ with respect to (U, φ) and (U', φ'), respectively. It follows from Eqn. (5) that

$$\sum_{i=1}^{n} \xi'^{i} \frac{\partial \tilde{f}'}{\partial x'^{i}}\bigg|_{\underline{x}'} = \xi(f) = \sum_{i=1}^{n} \xi^{i} \left[\frac{\partial \tilde{f}}{\partial x^{i}}\bigg|_{\underline{x}} \right]$$
$$= \sum_{i=1}^{n} \xi^{i} \left[\sum_{j=1}^{n} \frac{\partial \tilde{f}'}{\partial x'^{j}}\bigg|_{\underline{x}'} \frac{\partial x'^{j}}{\partial x^{i}}\bigg|_{\underline{x}} \right] \tag{6}$$

where $\underline{x} = \varphi(p)$ and $\underline{x}' = \varphi'(p)$. But Eqn. (6) must hold for all smooth f. Therefore,

$$\xi'^{1} = \sum_{i=1}^{n} \xi^{i} \frac{\partial x'^{1}}{\partial x^{i}}\bigg|_{\underline{x}}, \ldots, \quad \xi'^{n} = \sum_{i=1}^{n} \xi^{i} \frac{\partial x'^{n}}{\partial x^{i}}\bigg|_{\underline{x}} \tag{7}$$

[Exercise 25. Why does (7) follow from (6)?] Eqn. (7) describes the behavior of components (of a fixed vector) under change in choice of chart.

Exercise 26. Let p be a point of a manifold M. Suppose we are given a mapping from charts on M containing p, to n–tuples of real numbers, subject to (7). Show that such a mapping defines a contravariant vector in M at p.

Exercise 27. Show that, given a nonzero vector, ξ at p and an n-tuple of real numbers, $(\xi^1, \ldots, \xi^n)$, not all zero, there exists a chart with respect to which the components of ξ are $(\xi^1, \ldots, \xi^n)$.

4. Linear Algebra

In this section, we shall review some basic facts about linear algebra, and introduce some notation.

Let , V^a, V^b, ... be finite-dimensional vector spaces (over the reals). It is convenient to indicate to which V a vector belongs by means of superscript. Thus, an element of V^a would be written ξ^a, an element of V^c as η^c, etc. Hence, addition of two vectors is defined when and only when those vectors have the same superscript.

We define the dual of a vector space. Let $\mu : V^a \to R$ be linear, i.e., let μ be such that $\mu(\xi^a + m\eta^a) = \mu(\xi^a) + m\mu(\eta^a)$, for any number m and any ξ^a and η^a in V^a. If μ and ν are two such linear maps, and m is a real number, we define a new linear map, $(\mu + m\nu)$, by $(\mu + m\nu)(\xi^a) = \mu(\xi^a) + m\nu(\xi^a)$. With this definition of addition of linear maps, and multiplication of linear maps by (real) numbers, the linear maps from V^a to R form a vector space, called the dual of V^a. The dual of V^a will be written V_a, and membership in V_a will be indicated by the subscript "a". Thus, the μ above would be written μ_a. Finally, instead of $\mu(\xi^a)$, we write $\mu_a\xi^a$ or $\xi^a\mu_a$. It is clear that formulae such as

$$(\mu_a + m\nu_a)(\xi^a + k\eta^a) = \mu_a\xi^a + m\nu_a\xi^a + k\mu_a\eta^a + mk\nu_a\eta^a$$

are true. Similarly, the dual of V^b will be written V_b, and membership in V_b indicated by the subscript "b", etc. Thus, λ_c would represent an element of V_c, i.e., a linear mapping from V^c to R. In order that the sum of two vectors be defined, it is necessary that i) both vectors have subscripts or both vectors have superscripts, and ii) the subscripts or superscripts be the same letter.

We next define multilinear maps. Choose any finite (ordered) list of vector spaces, from the collection $V^a, V_a, V^b, V_b, \ldots$, having the following property: no vector space appears more than ones in the list, and no vector space appears together with its dual. (That is, no letter appears more than ones in the list.) A typical such list is (V^c, V_a, V_e, V^d). A mapping α from $V^c \times V_a \times V_e \times V^d$ (a product of sets) to R is said to be multilinear if α is linear in each variable separately, i.e., if

$$\alpha(\xi^c + m\eta^c, \mu_a, \lambda_e, \tau^d) = \alpha(\xi^c, \mu_a, \lambda_e, \tau^d) + m\alpha(\eta^c, \mu_a, \lambda_e, \tau^d),$$

$$\alpha(\xi^c, \mu_a + m\nu_a, \lambda_e, \tau^d) = \alpha(\xi^c, \mu_a, \lambda_e, \tau^d) + m\alpha(\xi^c, \nu_a, \lambda_e, \tau^a), \text{ etc.}$$

The collection of all such multilinear maps will be denoted by $V_c{}^{ae}{}_d$, and an element of $V_c{}^{ae}{}_d$ (e.g., α above) will be written $\alpha_c{}^{ae}{}_d$. Instead of $\alpha(\xi^c, \mu_a, \lambda_e, \tau^d)$, we write $\alpha_c{}^{ae}{}_d\xi^c\mu_a\lambda_e\tau^d$. Defining the sum of two multilinear mappings, and the product of a multilinear mapping with a number, by $(\alpha_c{}^{ae}{}_d + m\beta_c{}^{ae}{}_d)(\xi^c\mu_a\lambda_e\tau^d) = \alpha_c{}^{ae}{}_d\xi^c\mu_a\lambda_e\tau^d + m\beta_c{}^{ae}{}_d\xi^c\mu_a\lambda_e\tau^d$, we see that $V_c{}^{ae}{}_d$ has the structure of a vector space.

Thus, starting with the vector space $V^a, V^b, \dots$, we obtain a much larger collection of vector spaces, e.g., $V^{am}{}_b{}^c{}_{vk}$, etc. The elements of these vector spaces are called *tensors*. The superscripts attached to a tensor are called *contravariant indices*, the subscripts *covariant indices*. The *rank* of a tensor is defined as the total number of indices it possesses. Tensors of rank one are usually called *vectors*. Real numbers, when considered as tensors of rank zero, are called *scalars*.

What operations are available between these tensors? Of course, one can multiply a tensor by a scalar. This operation can be generalized. Consider two tensors which have no index letter in common (i.e., either "a" appears in precisely one of the tensors, or "a" appears in nether, etc. for "b", "c", etc....), e.g., consider $\alpha_m{}^e$ and $\beta^{ca}{}_b{}^p$. Then the expression $(\alpha_m{}^e\xi^m\eta_e)(\beta^{ca}{}_b{}^p\mu_c\nu_a\sigma^b\tau_p)$ evidently defines a multilinear map from $V^m \times V_e \times V_c \times V_a \times V^b \times V_p$ to R. That is, we have defined an element of $V_m{}^{eca}{}_b{}^p$. This element will be written $\alpha_m{}^e\beta^{ca}{}_b{}^p$. Thus, two tensors having the property that they share no index letter define a third tensor. This operation is called *outer product*. The rank of the outer product of two tensors is the sum of the ranks of the two tensors. Multiplication of a tensor by a scalar can be considered as a special case of the outer product.

Exercise 28. Check that $\alpha_n{}^e(\beta^{ca}\gamma_n{}^{pq}) = (\alpha_m{}^e\beta^{ca})\gamma_n{}^{pq}$. (Hence, parenthesis in such expressions are unnecessary.)

The second operation is addition. Of course, one can add two tensors having exactly the same index structure (i.e., having the same index letters, in the same order, and in the same locations), e.g., $\alpha_a{}^r{}_c + \beta_a{}^r{}_c$. It is convenient, however, to permit addition in more general situations. Evidently, the multilinear maps from $V^c \times V_n \times V_p$ to R are in one-to-one correspondence with multilinear maps from $V_n \times V_p \times V^c$ to R etc. That is to say, there are natural isomorphisms between the vector spaces $V_c{}^{np}$, $V^{np}{}_c$, $V^n{}_c{}^p$, $V^p{}_c{}^n$, $V^{pn}{}_c$, $V_c{}^{pn}$. Thus, an element of one of these vector spaces defines an element of each space. Whenever addition or equality is indicated between two tensors in isomorphic vector spaces, it is understood that one tensor is to be taken, using the isomorphism, into the vector space of the other, and there the sum is to be evaluated, or equality is to hold.

Example 7. The equation $\gamma^{np}{}_c = \alpha^n{}_c{}^p + \beta_c{}^{pn}$ makes the following statement. Using the isomorphism between $V_c{}^{pn}$ and $V^n{}_c{}^p$, take $\beta_c{}^{pn}$ (an element of V_c^{pn}) to $V^{n\ p}_{\ c}$. Add the result to $\alpha^{n\ p}_{\ c}$ (an element of $V^{n\ p}_{\ c}$). Using the isomorphism between $V^{n\ p}_{\ c}$ and $V^{np}_{\ \ c}$, take the resulting sum (an element of $V^{n\ p}_{\ c}$) to $V^{np}_{\ \ c}$. The result is the same as the element $\gamma^{np}_{\ \ c}$ of $V^{np}_{\ \ c}$. Thus, addition is defined between tensors having the property that they have exactly the same letters as contravariant indices and exactly the same letters as covariant indices.

Exercise 29. State and prove: $\xi^a_{\ p}\eta^{dm} = \eta^{dm}\xi^a_{\ p}$.

In fact, we have developed the formalism above in a context slightly more general that we shall need. Let V be a single finite-dimensional vector space, and let V^a, $V^b, \ldots,$ be copies of V. (That is to say, we are given isomorphisms between V^a and V, between V^b and V, etc.) Using the construction above, we obtain in this case, just as in the more general case, tensors. The same operations as above apply: the outer product is defined between any two tensors having no index letter in common: addition and equality are defined between any two tensors having no index letter in common; addition and equality are defined between any two tensors having exactly the same contravariant index letters and exactly the same covariant index letters. However, two additional operations are available in this special case.

Consider one of our vector space, e.g., $V^m_{\ \ ap}{}^d$. Choose two index letters, the first of which appears in $V^m_{\ \ ap}{}^d$, and the second of which does not, e.g., “p”and “b”. The isomorphism between V^p to V^b then, evidently, defines an isomorphism between $V^m_{\ \ ap}{}^d$ and $V^m_{\ \ ab}{}^d$. Thus, any element, e.g., $\alpha^m_{\ \ ap}{}^d$, of $V^m_{\ \ ap}{}^d$ defines some element of $V^m_{\ \ ab}{}^d$. This element of $V^m_{\ \ ab}{}^d$ will be written $\alpha^m_{\ \ ab}{}^d$. We say that $\alpha^m_{\ \ ab}{}^d$ has been obtained from $\alpha^m_{\ \ ap}{}^d$ by *index substitution.* Tensors, which are obtained from each other by index substitution are not written as equal, i.e., we do not write $\alpha^m_{\ \ ab}{}^d = \alpha^m_{\ \ ap}{}^d$. Index substitution is always indicated as above: one changes the letter which appears as an index, without changing the location of that index or the base letter. (“α” in the example above).

Example 8. Let $\beta_{mc}{}^d = \tau^d_{\ m}\mu_c$. Perform index substitution on $\beta_{mc}{}^d$ to obtain $\beta_{mc}{}^a$. Perform index substitution on $\tau^d_{\ m}$ to obtain $\tau^a_{\ m}$. Then $\beta_{mc}{}^a = \tau^a_{\ m}\mu_c$.

Example 9. Let β^{ab} be a tensor. Perform index substitution on β^{ab} to obtain β^{mb}; perform index substitution on β^{mb} to obtain β^{mn}; perform index substitution on β^{mn} to obtain β^{bn}; perform index substitution on β^{bn} to obtain β^{na}. Then, in general, it will not be true that $\beta^{ab} = \beta^{ba}$. If, however, ξ^a is a vector, ξ^b is the result of index substitution on ξ^a, and $\beta^{ab} = \xi^a\xi^b$, then it is true that $\beta^{ab} = \beta^{ba}$.

The second operation is called contraction. Consider a tensor with at least one covariant index and at least one contravariant index, e.g., $\alpha^{pbr}_{\ \ \ \ c}$. Choose two indices, one covariant and one contravariant, of $\alpha^{pbr}_{\ \ \ \ c}$, e.g.,“b” and “c”. This $\alpha^{pbr}_{\ \ \ \ c}$ can always be written as a sum of outer products of the form

$$\alpha^{pbr}{}_c = \lambda^{pr}\xi^b\eta_c + \cdots + \rho^{pr}\mu^b\nu_c$$

Now consider the tensor

$$\lambda^{pr}(\xi^b\eta_b) + \cdots + \rho^{pr}(\mu^b\nu_b)$$

an element of V^{pr} (the quantities $\xi^b\eta_b$, etc. are just scalars). This tensor (which is uniquely determined by $\alpha^{pbr}{}_c$, and the choice of indices "b" and "c") is called the *contraction* of $\alpha^{pbr}{}_c$ over the indices "b" and "c". It is written $\alpha^{pbr}{}_b$, or $\alpha^{pmr}{}_m$, etc., where the repeated index is any letter which does not appear elsewhere in α. When several contractions are applied to the same tensor, one uses different repeated indices to avoid confusion. Thus, contraction reduces the rank of a tensor by two. Simultaneous outer product and contraction are easily indicated by the notation, e.g., if $\alpha^{mb}{}_{ad} = \lambda^m{}_d\xi_a{}^b$, then $\alpha^{mc}{}_{ac}$ can be written $\lambda^m{}_c\xi_a{}^c$. The notations for the action of the dual (i.e., $\mu_a\xi^a$) and the action of a multilinear map (i.e., $\alpha^{mb}{}_a\xi_m\eta_b\lambda^a$) are clearly consistent with the notation for contraction, and, in fact we shall consider these actions as special cases of the contraction operation.

To summarize, given a single finite-dimensional vector space V, we obtain a collection of tensors, indexed by letters. On these tensors, there are defined four operations: i) outer product, defined between two tensors having no index letter in common, ii) addition, defined between two tensors having exactly the same contravariant index letters and exactly the same covariant index letters, iii) index substitution, which results in replacing one index letter of a tensor with any other letter which does appear elsewhere in the tensor, and iv) contraction, which results in replacing a pair of index letters, one contravariant and one covariant, by the same index letter which does not appear elsewhere in the tensor.

This notation is due to Penrose. Although the rules may sound rather formidable at first, one quickly gets used to them, and automatically performs the various operations where appropriate. The notation represents an important labor-saving device.

Example 10. The study of linear operators on a finite-dimensional vector space V is the study of tensors $\alpha^a{}_b$ over V. Composition of operators is defined by $\alpha^a{}_m\beta^m{}_b$, and the trace of an operator is $\alpha^a{}_a$.

Example 11. Let $\alpha^{ab} = \beta^{ab} + \beta^{ba}$. Then $\alpha^{ab} = \alpha^{ba}$. If $\gamma_{mn} = -\gamma_{nm}$, then $\alpha^{ab}\gamma_{ab} = 0$.

Example 12. If $\alpha^{ab} = \alpha^{ba}$. Then, for any ξ_p, $\xi_b\alpha^{ab} = \xi_d\alpha^{da}$.

Example 13. The following is meaningless: $\xi^{am}{}_{bc}{}^{da}$.

Example 14. If $\tau_b{}^{da} = \xi^a\mu_b{}^d + \eta^d\lambda_b{}^a$, then $\tau_m{}^{dm} = \xi^c\mu_c{}^d + \eta^d\lambda_e{}^e$

Example 15. Let $\delta^a{}_b$ be the (unique) element of $V^a{}_b$ having the property $\delta^a{}_b\xi^b = \xi^a$ for every ξ^b. Then $\delta^a{}_a$ = dimension of V. Furthermore, $\delta^a{}_c\alpha^{mb}{}_{ad} = \alpha^{mb}{}_{cd}$.

5. Tensors at a point

We have seen in Sect. 3 that the collection, $\mathfrak{F}'(p)$, of contravariant vectors at the point p of an n–dimensional manifold M form an n–dimensional vector space. We have seen in Sect. 4 that, given an n–dimensional vector space, one can construct tensors, e.g., $\lambda^{p}{}_{mcs}{}^{b}$, over that vector space. The tensors which result when this construction is applied to the vector space $\mathfrak{F}'\ (p)$ are called *tensors* in M at the point p. Between tensors in M at p, we have outer product, addition, index substitution, and contraction.

We illustrate by returning again to components. Let (U, φ) be a chart containing p, and let η_a be a covariant vector at p. Set $\eta_1 = \eta_a \xi^a$, where ξ^a is the contravariant vector at p with components $(1, 0 \ldots, 0)$; set $\eta_1 = \eta_a \mu^a$, where μ^a is the contravariant vector at p with components $(0, 1, 0, \ldots, 0)$, etc. The n numbers $(\eta_1, \eta_2, \ldots, \eta_b)$ are called the *components* of η_a with respect to the chart (U, φ). Evidently, if ξ^a is a contravariant vector at p with components $(\xi^1, \xi^2, \ldots, \xi^n)$, then $\eta_a \xi^a = \eta_1 \xi^1 + \cdots + \eta_n \xi^n$.

Now let (U', φ') be a second chart containing p. Let $\underline{x} = \varphi(p)$, $\underline{x}' = \varphi'(p)$, and let $x^1(x'^1, \ldots, x'^n), \ldots, x^n(x'^1, \ldots, x'^n)$ be the n functions of n variables defined by $\varphi \cdot \varphi'^{-1}$. Then, evidently, $\sum_{k=1}^{n} \frac{\partial x^i}{\partial x'^k}\big|_{\underline{x}'} \frac{\partial x'^k}{\partial x^j}\big|_{\underline{x}} = (1$ if $i = j, 0$, otherwise: $i, j, = 1, \ldots, n)$. Denote by $(\eta_1 \ldots, \eta_n)$ and $(\eta'_1 \ldots, \eta'_n)$ the components of η_a with respect to (U, φ) and (U, φ') respectively. Then

$$\sum_{i=1}^{n} \eta_i \xi^i = \eta_a \xi^a = \sum_{j=1}^{n} \eta'_j \xi'^j = \sum_{j=1}^{n} \sum_{i=1}^{n} \eta'_j \left. \frac{\partial x'^j}{\partial x^i} \right|_{\underline{x}} \xi^i$$

where we have used Eqn. (7). Evidently, this equation can hold for all ξ^a only if $\eta'_i = \sum_{j=1}^{n} \eta_j \frac{\partial x^j}{\partial x'^i}\big|_{\underline{x}'}$, $(i = 1, \ldots, n)$. This is the transformation law for the components of a (fixed) covariant vector.

Exercise 30. Define the components of an arbitrary tensor, e.g., $\alpha^a{}_{bc}$. Prove that these components have the following transformation law:

$$\alpha'^i{}_{jk} = \sum_{l=1}^{n} \sum_{p=1}^{n} \sum_{q=1}^{n} \alpha^l{}_{pq} \left. \frac{\partial x'^i}{\partial x^l} \right|_{\underline{x}} \left. \frac{\partial x^p}{\partial x'^j} \right|_{\underline{x}'} \left. \frac{\partial x^q}{\partial x'^k} \right|_{\underline{x}'}$$

$(i, j, k = 1, \ldots, n)$.

6. Tensor Fields

We now introduce the main objects of interest on a manifold: tensor fields.

Let M be a manifold. A *tensor field* on M consists of an assignment, to each point of M, a tensor at that point, where all the tensors at various points have the same index structure (i.e., the same index letters in the same locations). Thus, if $\alpha_m{}^{acn}{}_p$ is a tensor field on M, then, for each point p of M, $\alpha_m{}^{acn}{}_p(p)$ is a tensor at the point p.

Example 16. whereas the components of a tensor at a point consist of an array of real numbers, the components of a tensor field are functions of the n coordinates, $x^1, \ldots, x^n$

Evidently, the four tensor operations – addition, outer product, index substitution, and contraction – can be performed pointwise. Hence, these operations are also defined on tensor fields.

Example 17. Let $\alpha_m{}^{acn}{}_p$ be a tensor field. Then, for each point p $\alpha_m{}^{acn}{}_p(p)$ is a tensor at p. Since contraction is defined for tensors at p, we have, for each point p, a tensor $\alpha_d{}^{adn}{}_p(p)$. Thus, we have a tensor field, written $\alpha_d{}^{adn}{}_p$.

There is a somewhat more direct way of introducing tensor fields. Let ξ^a be a contravariant vector field on M, and let f be a smooth function on M. Then, for each point p of M, ξ^a (p) is a contravariant vector at p. But, by definition (of a contravariant vector, Sect. 3) $\xi^a(p)$ assigns, to each smooth function on M, a real number. In particular, ξ^a (p) assigns a real number to our smooth function f. But this is true at each point p, and so we obtain a real number assigned to each point of M. In other words, we obtain a function on M. We write this function $\mathcal{L}_\xi f$. Thus, given any contravariant vector field ξ^a on M and any smooth function f on M, $\mathcal{L}_\xi f$ is a function on M. We can regard $\mathcal{L}_\xi$ as a mapping from smooth function on M to functions on M. The conditions DD1, DD2, and DD3 (Sect. 3) become, respectively,

VF1. $\mathcal{L}_\xi(f+g) = \mathcal{L}_\xi f + \mathcal{L}_\xi g$.
VF2. $\mathcal{L}_\xi(fg) = f\mathcal{L}_\xi g + g\mathcal{L}_\xi f$.
VF3. If $f = \text{const.}$, then $\mathcal{L}_\xi f = 0$.

Thus, we could just as well have defined a vector field as a mapping, $\mathcal{L}_\xi$, from smooth functions on M to functions on M satisfying VF1, VF2, and

VF3.

Next, let μ_a be a covariant vector field on M. Then, if ξ^a is any contravariant vector field on M, $\mu_a\xi^a$ is a function on M. Thus, the covariant vector field μ_a defines a mapping, $F(\xi^a)$, from contravariant vector fields on M to functions on M. If ξ^a and η^a are contravariant vector fields on M, and f is any smooth function on M, then, evidently

$$F(\xi^a + f\eta^a) = F(\xi^a) + fF(\eta^a) \tag{8}$$

Now let F be any mapping from contravariant vector fields on M to functions on M, and suppose that F is linear in the sense of (8). Then, at each point p of M, F defines a mapping from contravariant vectors at p to numbers, and, by (8), that mapping is linear. In other words, F assigns, to each point p of M, F defines a mapping from contravariant vectors at p to numbers, and, by (8), that mapping is linear. In other words, F assigns, to each point p of M, a covariant vector at p. That is, F defines a covariant vector field on M. We have shown that there is a one-to-one correspondence between covariant vector fields on M and linear mapping (linear in the sense of (8)) from contravariant vector fields on M to functions on M. Thus, we could just as well have defined a covariant vector field as a linear mapping (in the sense of (8)) from contravariant vector fields on M to functions on M.

Finally, let $\alpha^a{}_{bc}$ be a tensor field on M. Then, if μ_a, ξ^b, τ^c are vector fields on M, $\alpha^a{}_{bc}\,\mu_a\,\xi^b\,\tau^c$ is a function on M. So, $\alpha^a{}_{bc}$ defines a mapping $F(\mu_a, \xi^b, \tau^c)$, from certain vector fields on M to functions on M. This mapping is multilinear, in the sense that, if f is any smooth function on M, then

$$\begin{aligned} F(\mu_a + f\nu_a, \xi^b, \tau^c) &= F(\mu_a, \xi^b, \tau^c) + fF(\nu_a, \xi^b, \tau^c) \\ F(\mu_a, \xi^b + f\eta^b, \tau^c) &= F(\mu_a, \xi^b, \tau^c) + fF(\mu_a, \eta^b, \tau^c) \\ F(\mu_a, \xi^b, \tau^c + f\sigma^c) &= F(\mu_a, \xi^b, \tau^c) + fF(\mu_a, \eta^b, \sigma^c) \end{aligned} \tag{9}$$

Conversely, a mapping $F(\mu_a, \xi^b, \tau^c)$ from certain vector fields on M to functions on M, multilinear in the sense of (9), defines a tensor field on M. Thus, we could just as well have defined the tensor fields on M as mapping, multilinear in the sense of (9), from vector fields on M to functions on M.

The purpose of the discussion above is to emphasize the following point: one could easily have defined the tensor fields directly without first defining tensors at a point. Contravariant vector fields can be defined in terms of their action (subject to VF1, VF2, and VF3) on smooth scalar fields; covariant vector fields in terms in their action (subject to (8)) on contravariant vector fields; tensor fields in terms of their action (subject to (9)) on vector fields. The discussions of Sections 3 and 5 can be regarded as motivation.

A contravariant vector field ξ^a is said to be *smooth* if, for every smooth function f, $\mathcal{L}_\xi f$ is a smooth function. A covariant vector field μ_a is said to be *smooth* if, for every smooth contravariant vector field ξ^a,$\mu_a\xi^a$ is a smooth function. A tensor field, e.g., $\alpha^a{}_{bc}$ is said to be *smooth* if, for any smooth vector fields $\mu_a, \xi^b, \tau^c, \alpha^a{}_{bc}\ \mu_a\tau^c$ is a smooth function. The smooth fields are the interesting ones. Hence, we shall hereafter normally omit the word "smooth": tensor fields are understood to be smooth unless otherwise stated. Smooth functions are sometimes called (smooth) *scalar fields*. It is clear from these definitions that the tensor operations – outer product, addition, index substitution, and contraction – produce smooth tensor fields when applied to smooth tensor fields.

Example 18. Let ξ^a be a (not necessarily smooth) contravariant vector field on M. Let (U, φ) be a chart. Let f be a smooth function on M, and set $\tilde{f} = f\cdot\varphi^{-1}, g = \mathscr{L}_\xi f$, and $\mathscr{L} = g\cdot\varphi^{-1}$. Let $\xi^1(x), \ldots, \xi^n(x)$ be the components of ξ^a with respect to this chart (Example 16). Then, from Eqn.5,

$$\tilde{g}(x) = \xi^1(x)\frac{\partial \tilde{f}}{\partial x^1} + \cdots + \xi^n(x)\frac{\partial \tilde{f}}{\partial x^n}$$

It follows immediately from this equation that a contravariant vector field is smooth if and only if its components with respect to every chart are C^∞ functions of $(x^1, \ldots, x^n)$.

Exercise 31. Prove that a covariant vector field is smooth if and only if its components with respect to every chart are C^∞ functions. Same for an arbitrary tensor field.

The pattern above, an important one, will be repeated on several later occasions. One starts with something about the scalar fields, then extends it to the contravariant vector fields using the action $\mathscr{L}_\xi f$, then extends it to the covariant vector fields using the action $\xi^a\mu_a$, and finally extends it to all the tensor fields using the action $\alpha^a{}_{bc}\mu_a\xi^b\tau^c$.

7. Lie Derivatives

Let ξ^a be a (smooth) contravariant vector field, f a (smooth) scalar field, and consider the scalar field $\mathcal{L}_\xi f$. Intuitively, ξ^a represents a "magnitude and direction" at each point of the manifold M. If, for each point of M, we take the "derivative of f" in the "direction of ξ^a", we obtain a a number. Doing so at each point, we obtain a scalar field on M, namely $\mathcal{L}_\xi f$. So, $\mathcal{L}_\xi f$ represents a sort of "directional derivative of f, at each point, in the direction of ξ^a". This $\mathcal{L}_\xi f$ is sometimes called the *Lie derivative* of f in the ξ^a–direction. Is it possible to define a "directional derivative in the ξ^a–direction" for other things, e.g., for tensor (in addition to scalar) fields? The answer is yes. We do so in this section.

Let ξ^a and η^a be vector fields, f a scalar field. Then $\mathcal{L}_\eta f$ is also a scalar field, so $\mathcal{L}_\xi \mathcal{L}_\eta f$ is still another scalar field. Set

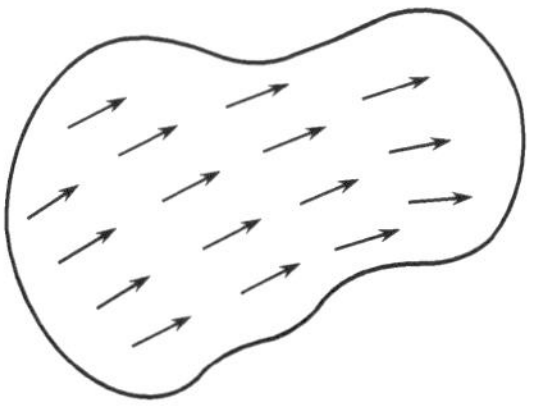

$$\kappa(f) = \mathcal{L}_\xi \mathcal{L}_\eta f - \mathcal{L}_\eta \mathcal{L}_\xi f \qquad (10)$$

so κ is a mapping from scalar fields (i.e., from $\mathfrak{F}$) to scalar fields. It follows immediately from VF1 that $\kappa(f + g) = \kappa(f) + \kappa(g)$. Furthermore, from VF1 and VF2,

$$\begin{aligned}
\kappa(fg) &= \mathcal{L}_\xi \mathcal{L}_\eta (fg) - \mathcal{L}_\eta \mathcal{L}_\xi (fg) \\
&= \mathcal{L}_\xi (f \mathcal{L}_\eta g + g \mathcal{L}_\eta f) - \mathcal{L}_\eta (f \mathcal{L}_\xi g + g \mathcal{L}_\xi f) \\
&= (\mathcal{L}_\xi f)(\mathcal{L}_\eta g) + f \mathcal{L}_\xi \mathcal{L}_\eta g + (\mathcal{L}_\xi g)(\mathcal{L}_\eta f) + g \mathcal{L}_\xi \mathcal{L}_\eta f \\
&- (\mathcal{L}_\eta f)(\mathcal{L}_\xi g) - f \mathcal{L}_\eta \mathcal{L}_\xi g - (\mathcal{L}_\eta g)(\mathcal{L}_\xi f) - g \mathcal{L}_\eta \mathcal{L}_\xi f \\
&= f\, k(g) + g\, k(f)
\end{aligned}$$

Finally, from VP3, $\kappa(f) = 0$ if $f = \text{const}$. We have just shown that $\kappa(f)$ itself satisfies the three conditions VF1, VF2, and VF3. Hence, by the remarks of Sect. 6 (on defining contravariant vector fields), we have $\kappa(f) = \mathcal{L}_\tau f$ (for all f) for some vector field τ^a. (Exercise 32. How do we know that τ^a is smooth?)

By definition, the *Lie derivative* of η^a in the ξ^a–direction, $\mathcal{L}_\xi \eta^a$, is the vector field τ^a. In other words, we define $\mathscr{L}_\xi \eta^a$ by the equation

$$\mathscr{L}_\xi \mathscr{L}_\eta f - \mathscr{L}_\eta \mathscr{L}_\xi f = \mathscr{L}_{(\mathscr{L}_\xi \eta)} f \tag{11}$$

for all f. The Lie derivative of a scalar field is a scalar field; the Lie derivative of a contravariant vector field is a contravariant vector field.

We have now extended the notion of the Lie derivative (in the ξ^a–direction) from scalar fields to contravariant vector fields. Let's work out some properties of it. First note that, for any scalar fields α and f, and vector fields ξ^a and λ^a,

$$\mathscr{L}_{(\alpha\xi+\lambda)} = \alpha \mathscr{L}_\xi f + \mathscr{L}_\lambda f \tag{12}$$

Applying (12), in the case $\alpha = 1$, to the definition (11), it follows immediately that

$$\mathscr{L}_\xi(\eta^a + \omega^a) = \mathscr{L}_\xi \eta^a + \mathscr{L}_\xi \omega^a \tag{13}$$

Now let α be a scalar field on M. What does $\mathscr{L}_\xi(\alpha\eta^a)$ look like? It's easy to find out from the definition (11):

$$\begin{aligned} \mathscr{L}_{(\mathscr{L}_\xi \alpha\eta)} f &= \mathscr{L}_\xi \mathscr{L}_{\alpha\eta} f - \mathscr{L}_{\alpha\eta} \mathscr{L}_\xi f \\ &= \mathscr{L}_\xi(\alpha \mathscr{L}_\eta f) - \alpha \mathscr{L}_\eta \mathscr{L}_\xi f \\ &= (\mathscr{L}_\xi \alpha)(\mathscr{L}_\eta f) + \alpha \mathscr{L}_\xi \mathscr{L}_\eta f - \alpha \mathscr{L}_\eta \mathscr{L}_\xi f \\ &= \mathscr{L}_{(\eta \mathscr{L}_\xi \alpha)} f + \alpha \mathscr{L}_{(\mathscr{L}_\xi \eta)} f \\ &= \mathscr{L}_{(\eta \mathscr{L}_\xi \alpha + \alpha \mathscr{L}_\xi \eta)} f \end{aligned} \tag{14}$$

Hence, since f is arbitrary,

$$\mathscr{L}_\xi(\alpha\eta^a) = \eta^a \mathscr{L}_\xi \alpha + \alpha \mathscr{L}_\xi \eta^a \tag{15}$$

(Compare Eqn. (13) with VF1: Eqn. (15) with VP2.)

The next step is to define Lie derivatives of covariant vector fields. Let μ^a be a covariant vector field, and consider the scalar field

$$\mathscr{L}_\xi(\eta^a \mu_a) - \mu_a \mathscr{L}_\xi \eta^a \tag{16}$$

where η^a is any contravariant vector field. Replacing η^a in (16) by $\eta^a + \alpha\omega^a$ (α a scalar field),

$$\begin{aligned} &\mathscr{L}_\xi[(\eta^a + \alpha\omega^a)\mu_a] - \mu_a \mathscr{L}_\xi(\eta^a + \alpha\omega^a) \\ &= \mathscr{L}_\xi(\eta^a \mu_a) + \mathscr{L}_\xi[\alpha(\omega^a \mu_a)] - \mu_a \mathscr{L}_\xi \eta^a - \mu_a \mathscr{L}_\xi(\alpha\omega^a) \\ &= \mathscr{L}_\xi(\eta^a \mu_a) + \alpha \mathscr{L}_\xi(\omega^a \mu_a) + (\omega^a \mu_a)\mathscr{L}_\xi \alpha - \mu_a \mathscr{L}_\xi \eta^a \\ &- \alpha\mu_a \mathscr{L}_\xi \omega^a - (\mu_a \omega^a)\mathscr{L}_\xi \alpha \\ &= [\mathscr{L}_\xi(\eta^a \mu_a) - \mu_a \mathscr{L}_\xi \eta^a] + \alpha[\mathscr{L}_\xi(\omega^a \mu_a) - \mu_a \mathscr{L}_\xi \omega^a] \end{aligned}$$

where we have used (13), (15) and VF2. In other words, the expression (16) defines a linear (in the sense of (8)) mapping from contravariant vector fields (η^a) to scalar fields on M. By the remarks of Sect. 6 (on the definition of a covariant vector field), such a mapping represents a covariant vector field. We define the *Lie derivative* of μ_a (in the ξ^a–direction), $\mathscr{L}_\xi \mu_a$, to be this vector field. That is, we define $\mathscr{L}_\xi \mu_a$ by the equation

$$\eta^a \mathscr{L}_\xi \mu_a = \mathscr{L}_\xi(\eta^a \mu_a) - \mu_a \mathscr{L}_\xi \eta^a \tag{17}$$

for all η^a.

We have now extended the notion of Lie derivative to covariant fields (the result again being a covariant vector field). Again, we want to work out some properties. It follows immediately from VF1, Eqn. (13), and the definition (17) that

$$\mathscr{L}_\xi(\mu_a + \nu_a) = \mathscr{L}_\xi \mu_a + \mathscr{L}_\xi \nu_a \tag{18}$$

To evaluate $\mathscr{L}_\xi(\alpha\mu_a)$, we use VF2 and the definition (17):

$$\begin{aligned} \eta^a \mathscr{L}_\xi(\alpha\mu_a) &= \mathscr{L}_\xi(\alpha\mu_a\eta^a) - \alpha\mu_a \mathscr{L}_\xi \eta^a \\ &= \mu_a\eta^a \mathscr{L}_\xi \alpha + \alpha \mathscr{L}_\xi(\mu_a\eta^a) - \alpha\mu_a \mathscr{L}_\xi \eta^a \\ &= \eta^a[\mu_a \mathscr{L}_\xi \alpha + \alpha \mathscr{L}_\xi \mu_a] \end{aligned}$$

Since this equation must hold for all η^a,

$$\mathscr{L}_\xi(\alpha\mu_a) = \mu_a \mathscr{L}_\xi \alpha + \alpha \mathscr{L}_\xi \mu_a \tag{19}$$

The last step is to extend the notion of the Lie derivative from scalar and vector fields to arbitrary tensor fields. The method is the same. We merely do an example. Let $\alpha^a{}_{bc}$ be a tensor field, and consider the expression

$$\mathscr{L}_\xi(\alpha^a{}_{bc}\mu_a\eta^b\lambda^c) - \alpha^a{}_{bc}\eta^b\lambda^c \mathscr{L}_\xi \mu_a - \alpha^a{}_{bc}\mu_a\lambda^c \mathscr{L}_\xi \eta^b - \alpha^a{}_{bc}\mu_a\eta^b \mathscr{L}_\xi \lambda^c \tag{20}$$

Repeating (essentially unchanged) the calculation at the top of the preceding page, we see that the expression (20) is multilinear in μ_a, η^b, λ^c (in the sense of (9)). Hence, by the discussion of Sect. 6 (concerning definitions of tensor fields) the expression (20) defines a tensor field on M, which we denote by $\mathscr{L}_\xi \alpha^a{}_{bc}$. That is to say, we define $\mathscr{L}_\xi \alpha^a{}_{bc}$ by

$$\begin{aligned} \mu_a\eta^b\lambda^c \mathscr{L}_\xi \alpha^a{}_{bc} &= \mathscr{L}_\xi(\alpha^a{}_{bc}\mu_a\eta^b\lambda^c) \\ &- \alpha^a{}_{bc}\eta^b\lambda^c \mathscr{L}_\xi \mu_a - \alpha^a{}_{bc}\mu_a\lambda^c \mathscr{L}_\xi \eta^b - \alpha^a{}_{bc}\mu_a\eta^b \mathscr{L}_\xi \lambda^c \end{aligned} \tag{21}$$

This is the *Lie derivative* of an arbitrary tensor field (in the ξ^a–direction). Linearity of the Lie derivative, i.e., the fact that

$$\mathscr{L}_\xi(\alpha^a{}_{bc} + \beta^a{}_{bc}) = \mathscr{L}_\xi \alpha^a{}_{bc} + \mathscr{L}_\xi \beta^a{}_{bc} \tag{22}$$

follows immediately from VF1.

Eqns. (15) and (19) are suggestive. They can generalized. Let $\alpha^a{}_{bc} = \beta^a{}_b\gamma_c$ in (21). Then

$$\begin{aligned}
&\mu_a\eta^b\lambda^c\mathscr{L}_\xi(\beta^a{}_b\gamma_c) = \mathscr{L}_\xi(\beta^a{}_b\gamma_c\mu_a\eta^b\lambda^c)\\
&-\beta^a{}_b\gamma_c\eta^b\lambda^c\mathscr{L}_\xi\mu_a - \beta^a{}_b\gamma_c\mu_a\lambda^c\mathscr{L}_\xi\eta^b - \beta^a{}_b\gamma_c\mu_a\eta^b\mathscr{L}_\xi\lambda^c\\
&= \beta^a{}_b\mu_a\eta^b\mathscr{L}_\xi(\gamma_c\lambda^c) + \gamma_c\lambda^c\mathscr{L}_\xi(\beta^a{}_b\mu_a\eta^b)\\
&-\beta^a{}_b\gamma_c\eta^b\lambda^c\mathscr{L}_\xi\mu_a - \beta^a{}_b\gamma_c\eta_a\lambda^c\mathscr{L}_\xi\eta^b - \beta^a{}_b\gamma_c\mu_a\eta^b\mathscr{L}_\xi\lambda^c\\
&= (\beta^a{}_b\mu_a\eta^b)(\lambda^c\mathscr{L}_\xi\gamma_c) + (\gamma_c\lambda^c)(\mu_a\eta^b\mathscr{L}_\xi\beta^a{}_b)
\end{aligned}$$

Since μ_a,η^b, and λ^c are arbitrary, we have

$$\mathscr{L}_\xi\beta^a{}_b\gamma_c = \beta^a{}_b\mathscr{L}_\xi\gamma^c + \gamma_c\mathscr{L}_\xi\beta^a{}_b \tag{23}$$

A similar equation holds, of course, for arbitrary outer products. Note that Eqns. (15) and (19) (as well as VF2) are special cases in which one factor in the outer products is a scalar field. Finally, rewriting Eqn. (17) in the form $\mathscr{L}_\xi(\mu_a\eta^a) = \mu_a\mathscr{L}_\xi\eta^a + \eta^a\mathscr{L}_\xi\mu_a$, and noting the definition of contraction, we see that the contraction of the Lie derivative of any tensor field is equal to the Lie derivative of its contraction.

We summarize the properties of the Lie derivative in the following:

Theorem 8. Let ξ^a be a contravariant vector field on the manifold M. Then with any tensor field, $\alpha^{a...c}{}_{b...d}$, on M, there is associated a second tensor field, $\mathscr{L}_\xi\alpha^{a...c}{}_{b...d}$, where this operation on tensor fields has the following properties:

LD1. $\mathscr{L}_\xi(\alpha^{a...c}{}_{b...d} + \beta^{a...c}{}_{b...d} = \mathscr{L}_\xi(\alpha^{a...c}{}_{b...d}) + \mathscr{L}_\xi(\beta^{a...c}{}_{b...d})$.

LD2. $\mathscr{L}_\xi(\alpha^{a...c}{}_{b...d}\gamma^{m...n}{}_{p...q}) = \alpha^{a...c}{}_{b...d}\mathscr{L}_\xi(\gamma^{m...n}{}_{p...q})$ $+\gamma^{m...n}{}_{p...q}\mathscr{L}_\xi(\alpha^{a...c}{}_{b...d})$.

LD3. The operations $\mathscr{L}_\xi$ and index substitution can be performed in either order, with same result.

LD4. The operations $\mathscr{L}_\xi$ and construction can be performed in either order, with the same result.

Property LD2 is sometimes called the Leibnitz rule. That there are four properties in Theorem 4 is no coincidence: there are four fundamental operations on tensors (addition, outer product, index substitution, and contraction), so the properties above express the interaction between Lie derivatives and these tensor operations. Note that practically every interesting equation in this section is a consequence of the four properties above, e.g., VF1, VF2, (13), (15), (17), (18), (19), (21), and (23). Although the definition of the Lie derivative is, of course, worth knowing, it is the properties above which are most useful.

Two further remarks about the Lie derivative should be noted. Since Lie derivatives are always in the direction of a contravariant vector field, the Lie derivatives of contravariant vector fields have an especially pretty structure. Let ξ^a and η^a be contravariant vector fields, and, for the purposes of this paragraph, set $[\xi, \eta] = \mathscr{L}_\xi \eta^a$. Then, from (11), we have immediately

$$[\xi, \eta] = -[\eta, \xi] \tag{24}$$

Furthermore, we have, for three contravariant vector fields,

$$[\lambda, [\xi, \eta]] + [\xi, [\eta, \lambda]] + [\eta, [\lambda, \xi]] = 0 \tag{25}$$

(Proof: Apply the left side of (25) to an arbitrary scalar field f. Now expand, using (11), and note that all the terms cancel.) So, on the contravariant vector fields on a manifold, there is defined a skew (Eqn. (24)) "bracket" which obeys Jacobi's identity (Eqn. (25)). That is to say, the contravariant vector fields on a manifold have the structure of a Lie algebra.

The second remark concerns the dependence of the Lie derivative on the vector field representing the direction in which the Lie derivative is taken. (Theorem 8 refers to the dependence on the tensor field which is "Lie derived".) We need a bit of notation. Fix a scalar field α. Then, foe each ξ^a, $\mathscr{L}_\xi \alpha$ is a scalar field. By (12), this scalar field is linear (in the sense of (8)) in ξ^a. Hence, α defines a covariant vector field. This vector field, which is called the *gradient* of α, is written $D(\alpha)_a$. Thus, $D(\alpha)_a$ is defined by

$$\xi^a D(\alpha)_a = \mathscr{L}_\xi \alpha \tag{26}$$

For Lie derivatives of scalar fields, the dependence on ξ^a is given by (12). For contravariant vector fields, the dependence follows immediately from Eqn. (15) and the fact that $\mathscr{L}_\xi \eta^a = \mathscr{L}_\eta \xi^a$:

$$\mathscr{L}_{(\alpha\xi+\lambda)} \eta^a = \alpha \mathscr{L}_\xi \eta^a + \mathscr{L}_\lambda \eta^a - \xi^a \eta^m D(\alpha)_m \tag{27}$$

For covariant vector fields, we use Eqn. (17):

$$\begin{aligned}
\tau^a \mathscr{L}_{(\alpha\xi+\lambda)} \mu_a &= \mathscr{L}_{(\alpha\xi+\lambda)} (\tau^a \mu_a) - \mu_a \mathscr{L}_{(\alpha\xi+\lambda)} \tau^a \\
&= \alpha \mathscr{L}_\xi (\tau^a \mu_a) + \mathscr{L}_\lambda (\tau^a \mu_a) - \mu_a \alpha \mathscr{L}_\xi \tau^a - \mu_a \mathscr{L}_\lambda \tau^a + \mu_a \xi^a \tau^b D(\alpha)_b \\
&= \tau^a \alpha \mathscr{L}_\xi \mu_a + \tau^a \mathscr{L}_\lambda \mu_a + \mu_a \xi^a \tau^b D(\alpha)_b
\end{aligned}$$

Since this holds for all τ^a, we have

$$\mathscr{L}_{(\alpha\xi+\lambda)} \mu_a = \alpha \mathscr{L}_\xi \mu_a + \mathscr{L}_\lambda \mu_a + D(\alpha)_a \mu_b \xi^b \tag{28}$$

Exercise 33. Prove that, for an arbitrary tensor field, e.g., $\beta^a{}_{bc}$, we have

$$\begin{aligned}
\mathscr{L}_{(\alpha\xi+\lambda)} \beta^a{}_{bc} &= \alpha \mathscr{L}_\xi \beta^a{}_{bc} + \mathscr{L}_\lambda \beta^a{}_{bc} - \xi^a \beta^m{}_{bc} D(\alpha)_m \\
&\quad + D(\alpha)_b \beta^a{}_{mc} \xi^m + D(\alpha)_c \beta^a{}_{bm} \xi^m
\end{aligned} \tag{29}$$

(Hint: Use (27), (28), and (21).) Note that (12), (27), and (28) are special cases of (29).

Exercise 34. Express Lie derivatives in terms of components.

Example 19. Let ξ^a and μ_a be vector fields. Then $\xi^a\mu_a$ is a scalar field. Hence, we have a multilinear mapping (in the sense of (9)) from smooth ξ^a, μ_a to smooth scalar fields. So we have defined a tensor field, which we write $\delta^a{}_b$ (See Example 15.) That is, $\delta^a{}_b$ is defined by $\delta^a{}_b\mu_a\xi^b = \mu_a\xi^a$, for every μ_a, ξ^b. (This $\delta^a{}_b$ is called the *unit tensor.*) We evaluate its Lie derivative. From Eqn. (21).

$$\begin{aligned}\alpha_a\beta^b\,\mathcal{L}_\xi\delta^a{}_b &= \mathcal{L}_\xi(\delta^a{}_b\alpha_a\beta^b) - \delta^a{}_b\beta^b\,\mathcal{L}_\xi\alpha_a - \delta^a{}_b\alpha_a\mathcal{L}_\xi\beta^b\\ &= \mathcal{L}_\xi(\alpha_a\beta^a) - \beta^a\,\mathcal{L}_\xi\alpha_a - \alpha\,\mathcal{L}_\xi\beta^a\end{aligned}$$

But now, from Eqn, (17), we have $\mathcal{L}_\xi\delta^a{}_b = 0$.

Exercise 35. Find the components of $\delta^a{}_b$ with respect to a chart, and verify the conclusion of Example 19 using your formula of Exercise 34.

Exercise 36. A constant scalar field has the property that its Lie derivative in the direction of every vector field is zero. Thus, one might be tempted to define a "constant tensor field" as one whose Lie derivative in the direction of every vector field is zero. Show that this idea does not work, in the following sense: a "constant tensor field" of rank one necessarily vanishes.

Example 20. Let ξ^a be a vector field, α a scalar field. We prove that

$$D(\mathcal{L}_\xi\alpha)_a = \mathcal{L}_\xi D(\alpha)_a \tag{30}$$

To do this note that

$$\begin{aligned}\eta^a\,\mathcal{L}_\xi D(\alpha)_a &= \mathcal{L}_\xi\eta^a D(\alpha)_a - D(\alpha)_a\mathcal{L}_\xi\eta^a = \mathcal{L}_\xi\mathcal{L}_\eta\alpha - \mathcal{L}_{(\mathcal{L}_\xi\eta)}\alpha\\ &= \mathcal{L}_\eta\mathcal{L}_\xi\alpha = \eta^a D(\mathcal{L}_\xi\alpha)_a\end{aligned}$$

where we have used (11). Since η^a is arbitrary, (30) follows. (That is, the Lie derivative commutes with the gradient.) Note that Eqn. (30) implies Eqn. (11).

Exercise 37. Can $\mathcal{L}_\xi$ be defined in any reasonable way when ξ^a is not smooth?

Exercise 38. Find a list of axioms on an operator K on tensors which are necessary and sufficient for $K = \mathcal{L}_\eta$ for some vector field ξ^a.

Exercise 38. Let ξ^a and η^a be contravariant vector fields. Prove that

$$\mathcal{L}_\xi\mathcal{L}_\eta\alpha^{a\ldots c}{}_{b\ldots d} - \mathcal{L}_\eta\mathcal{L}_\xi\alpha^{a\ldots c}{}_{b\ldots d} = \mathcal{L}_{(\mathcal{L}_\xi\eta)}\alpha^{a\ldots c}{}_{b\ldots d} \tag{31}$$

(This equation generalizes (11).) (For scalars, (31) follows from (11). For contravariant vectors, contract (31) with $D(f)_a$, f arbitrary, and verify using

(11). Now extend to covariant vectors, and finally to tensors of arbitrary rank.)

We shall obtain a more geometrical interpretation of the Lie derivative later.

8. Symmetrization and Antisymmetrization

Let $\alpha_{a\ bc}^{\ m\ \ d}$ and $\beta^{mn\ cb}_{\ \ a}$ be tensor fields. Set

$$\alpha_{(a\ b)c}^{\ \ m\ \ \ d} = \frac{1}{2}(\alpha_{a\ bc}^{\ m\ \ d} + \alpha_{b\ ac}^{\ m\ \ d})$$

$$\alpha_{[a\ b]c}^{\ \ m\ \ \ d} = \frac{1}{2}(\alpha_{a\ bc}^{\ m\ \ d} - \alpha_{b\ ac}^{\ m\ \ d})$$

$$\beta^{m(n\ cb)}_{\ \ \ a} = \frac{1}{6}(\beta^{mn\ cb}_{\ \ a} + \beta^{mc\ bn}_{\ \ a} + \beta^{mb\ nc}_{\ \ a} + \beta^{mn\ bc}_{\ \ a} + \beta^{mc\ nb}_{\ \ a} + \beta^{mb\ cn}_{\ \ a})$$

$$\beta^{m[n\ cb]}_{\ \ \ a} = \frac{1}{6}(\beta^{mn\ cb}_{\ \ a} + \beta^{mc\ bn}_{\ \ a} + \beta^{mb\ nc}_{\ \ a} - \beta^{mn\ bc}_{\ \ a} - \beta^{mc\ nb}_{\ \ a} - \beta^{mb\ cn}_{\ \ a})$$

More generally, a tensor, written with round brackets surrounding a collection of p consecutive indices (all covariant or all contravariant), means $1/p!$ times the sum of the $p!$ tensors obtained (using index substitution) by arranging those p indices in all possible orders. A tensor, written with square brackets surrounding a collection of p consecutive indices (all covariant or all contravariant), means $1/p!$ times a linear combination of the $p!$ tensors obtained by arranging those p indices in all possible orders, where a plus sign appears before those tensors in which the arrangement of indices is an even permutation of the original index arrangement, and a minus sign appears otherwise. These operations are called *symmetrization* and *antisymmetrization* (over the indices in question), respectively. These definitions occasionally make it possible to write expressions more concisely.

Example 21. $\tau_{[bac]} = -\tau_{[abc]} = -\tau_{[bca]} = \tau_{[cba]}$. $\tau_{(bac)} = \tau_{(abc)} = \tau_{(bca)} = \tau_{(cba)}$

Example 22. Set $\alpha_{abcdef} = \beta_{ab[cde]f}$. Then $\alpha_{a[bcdef]} = \beta_{a[bcdef]}$, while $\alpha_{a[bcdef]} = 0$

Example 23. $\xi^m \alpha_{[ma}\beta_{b]} = 2/3\xi^m \alpha_{m[a}\beta_{b]} + 1/3\alpha_{ab}\xi^m \beta_m$.

Example 24. If $\tau_{bac} = -\tau_{abc}$, $\tau_{acb} = -\tau_{abc}$, and $\tau_{cba} = -\tau_{abc}$, then $\tau_{abc} = \tau_{[abc]}$. More generally, $\alpha_{a\ldots c} = \alpha_{[a\ldots c]}$ if and only if reverses sign under interchange of any two of its indices.

Example 25. $\beta_{a\ldots c} = \beta_{(a\ldots c)}$ if and only if is invariant under interchange of any two of its indices.

Example 26. If $\xi^a\eta^b\alpha_{ab} = -\xi^b\eta^a\alpha_{ab}$ for all ξ^a, η^a then $(\alpha_{ab} + \alpha_{bc})\xi^a\eta^b = 0$ for all ξ^a, η^a, whence $(\alpha_{ab} + \alpha_{ab}) = 0$, whence $\alpha_{ab} = \alpha_{[ab]}$.

9. Exterior Derivatives

We now introduce a second type of "natural derivative" on manifolds. We summarize its structure:

	Lie Derivative	*Exterior Derivative*
Generalization of	Directional Derivative	Gradient
Requires	A Contravariant Vector Field	Nothing
Applicable to	All Tensor Fields	Tensor Fields $\omega_{a...c} = \omega_{[a...c]}$
Index Structure of Result	Same as Original Tensor Field	Adds One Index

Thus, the exterior derivative has the advantage over the Lie derivative that no extraneous tensor fields (i.e., ξ^a) are required, and the disadvantage that it is applicable only to tensor fields with a certain index structure. We shall see later that Lie derivatives and exterior derivatives are special cases (by far the most common and most useful ones) of a more general class of objects.

The *exterior derivative* of a scalar field, ω, is defined as its gradient, $D(\omega)_m$. (See Eqn. (30).) It follows immediately from VF2 that

$$D(\mu\omega)_m = \omega D(\mu)_m + \mu D(\omega)_m \tag{32}$$

where μ is any scalar field.

Now let ω_a be a contravariant vector field. Set $\lambda = \xi^a \omega_a$, and consider the right side of

$$\xi^m D(\omega)_{ma} = \frac{1}{2}\mathscr{L}_\xi \omega_a = \frac{1}{2}D(\lambda)_a \tag{33}$$

Replacing ξ^a by $\alpha\xi^a$ in the right side of (33), we have

$$\frac{1}{2}\mathcal{L}_{\alpha\xi}\omega_a - \frac{1}{2}D(x\lambda)_a = \frac{1}{2}\alpha\mathcal{L}_\xi\omega_a + \frac{1}{2}D(\alpha)_a\xi^m\omega_m - \frac{1}{2}\alpha D(\lambda)_a - \frac{1}{2}\lambda D(\alpha)_a$$
$$= \alpha\left[\frac{1}{2}\mathcal{L}_\xi\omega_a - D(\lambda)_a\right]$$

were we have used (32) and (28). Hence, Eqn. (33) defines a tensor field, $D(\omega)_{ma}$. This $D(\omega)_{ma}$ is the *exterior derivative* of the covariant vector field ω_a. To show that $D(\omega)_{ma}$ is antisymmetric, contract (33) with an arbitrary η^a:

$$\eta^a\xi^m D(\omega)_{ma} = \frac{1}{2}\eta^a\mathcal{L}_\xi\omega_a - \frac{1}{2}\eta^a D(\lambda)_a$$
$$= \frac{1}{2}\mathcal{L}_\xi(\eta^a\omega_a) - \frac{1}{2}\omega_a\mathcal{L}_\xi\eta^a - \frac{1}{2}\mathcal{L}_\eta(\xi^a\omega_a)$$

and note that the last expression above reverses sign under interchange of ξ^a and η^a. (The first and third terms together reverse sign, and the second term alone reverses sign.) Since ξ^a and η^a are arbitrary,

$$D(\omega)_{ma} = D(\omega)_{[ma]} \tag{34}$$

Finally, replacing ω_a by $\mu\omega_a$ in (33), we have

$$\xi^m D(\mu\omega)_{ma} = \frac{1}{2}\mathcal{L}_\xi(\mu\omega_a) - \frac{1}{2}D(\mu\lambda)_a$$
$$= \frac{1}{2}\mu\mathcal{L}_\xi\omega_a + \frac{1}{2}\omega_a\mathcal{L}_\xi\mu - \frac{1}{2}\lambda D(\mu)_a - \frac{1}{2}\mu D(\lambda)_a$$
$$= \mu\xi^m D(\omega)_{ma} + \frac{1}{2}\xi^m(D(\mu)_m\omega_a - \omega_m D(\mu)_a)$$

Since ξ^m is arbitrary,

$$D(\mu\omega)_{ma} = \mu D(\mu)_{ma} + D(\mu)_{[m}\omega_{a]} \tag{35}$$

Now let $\omega_{ab} = \omega_{[ab]}$ be a tensor field. Set $\lambda_b = \xi^a\omega_{ab}$, and consider the right side of

$$\xi^m D(\omega)_{mab} = \frac{1}{3}\mathcal{L}_\xi\omega_{ab} - \frac{2}{3}D(\lambda)_{ab} \tag{36}$$

Replacing ξ^m by $\alpha\xi^m$ in the right side of (36),

$$\frac{1}{3}\mathcal{L}_{\alpha\xi}\omega_{ab} - \frac{2}{3}D(\alpha\lambda)_{ab} = \frac{1}{3}\alpha\mathcal{L}_\xi\omega_{ab} + \frac{1}{3}D(\alpha)_a\xi^m\omega_{mb} + \frac{1}{3}D(\alpha)_b\xi^m\omega_{am}$$
$$- \frac{2}{3}\alpha D(\lambda)_{ab} - \frac{2}{3}D(\alpha)_{[a}\lambda_{b]} = \alpha\left[\frac{1}{3}\mathcal{L}_\xi\omega_{ab} - \frac{2}{3}D(\lambda)_{ab}\right]$$

Hence, Eqn. (36) defines a tensor field, $D(\omega)_{mab}$, which we call the *exterior derivative* of ω_{ab}. Contracting (36) with η^a,

$$\begin{aligned}\eta^a \xi^m D(\omega)_{mab} &= \frac{1}{3}\eta^a \mathcal{L}_\xi \omega_{ab} - \frac{2}{3}\eta^a D(\lambda)_{ab} \\ &= \frac{1}{3}\mathcal{L}_\xi \eta^a \omega_{ab} - \frac{1}{3}\omega_{ab}\mathcal{L}_\xi \eta^a - \frac{2}{3}\Big(\frac{1}{2}\mathcal{L}_\eta \lambda_b - \frac{1}{2}D(\kappa)_b\Big) \\ &= \frac{1}{3}\mathcal{L}_\xi(\eta^a \omega_{ab}) - \frac{1}{3}\mathcal{L}_\eta(\xi^a \omega_{ab}) - \frac{1}{3}\omega_{ab}\mathcal{L}_\xi \eta^a + \frac{1}{3}D(\kappa)_b\end{aligned}$$

where we have set $\kappa = \xi^a \eta^b \omega_{ab}$. Reversing the roles of ξ^a and η^a, the first two terms in the last expression above reverse sign. Since, furthermore, κ reverses sign, the entire expression reverses sign. Hence, since ξ^a and η^a are arbitrary, $D(\omega)_{mab}$ is antisymmetric in m and a. Since, furthermore, $D(\omega)_{mab}$ is (clearly, from(36)) antisymmetric in a and b, we have

$$D(\omega)_{mab} = D(\omega)_{[mab]} \tag{37}$$

Finally, replacing ω_{ab} in (36) by $\mu\omega_{ab}$,

$$\begin{aligned}\xi^m D(\mu\omega)_{mab} &= \frac{1}{3}\mathcal{L}_\xi(\mu\omega_{ab}) - \frac{2}{3}D(\mu\lambda)_{ab} \\ &= \mu\Big(\frac{1}{2}\mathcal{L}_\xi \omega_{ab} - \frac{2}{3}D(\lambda)_{ab}\Big) + \frac{1}{3}\omega_{ab}\mathcal{L}_\xi \mu - \frac{2}{3}D(\mu)_{[a}\lambda_{b]} \\ &= \mu D(\omega)_{mab}\xi^m + \xi^m D(\mu)_{[m}\omega_{ab]}\end{aligned}$$

Since ξ^m is arbitrary, we have

$$D(\mu\omega)_{mab} = \mu D(\omega)_{[m}\omega_{ab]} \tag{38}$$

We proceed by induction (on the rank of ω). Let $\omega_{a\ldots c} = \omega_{[a\ldots c]}$ have rank p. Define the *exterior derivative* of $\omega_{a\ldots c}$ by

$$\xi^m D(\omega)_{ma\ldots c} = \frac{1}{p+1}\mathcal{L}_\xi \omega_{a\ldots c} - \frac{p}{p+1}D(\lambda)_{a\ldots c}\lambda_{b\ldots c} = \xi^a \omega_{ab\ldots c} \tag{39}$$

Then, as above, one shows that

$$D(\omega)_{ma\ldots c} = D(\omega)_{[ma\ldots c]} \tag{40}$$

$$D(\mu\omega)_{ma\ldots c} = D(\mu)_{[m}\omega_{a\ldots c]} \tag{41}$$

A tensor field with p covariant indices and no contravariant indices, which is antisymmetric in all its indices, is called a *p-form*. Thus, the exterior derivative of a p-form is a (p+1)-form.

We next establish two further properties of the exterior derivative. Let $\omega_{a\ldots c}$ be a p-form. Then

$$\mathscr{L}_\xi D(\omega)_{ma\ldots c} = D(\mathscr{L}_\xi \omega)_{ma\ldots c} \tag{42}$$

$$D(D(\omega))_{mna\ldots c} = 0 \tag{43}$$

where, in (42), ξ^a is arbitrary. Eqn. (42) states that exterior derivatives and Lie derivatives commute. Eqn. (43) states that, if exterior differentiation is applied twice in succession, the result is zero. (The exterior derivative is often called the curl.) To prove (42) and (43), we again proceed by induction on p. First note (Example 30) that (42) holds when ω is a scalar field. Using this fact, we have, for ω a scalar field,

$$\begin{aligned}\xi^m D(D(\omega))_{ma} &= \frac{1}{2}\mathscr{L}_\xi D(\omega)_a - \frac{1}{2}D(\lambda)_a \\ &= \frac{1}{2}D(\mathscr{L}_\xi \omega)_a - \frac{1}{2}D(\lambda)_a = 0\end{aligned}$$

where $\lambda = \xi^a D(\omega)_a = \mathscr{L}_\xi \omega$. Hence, since ξ^m is arbitrary, (43) is satisfied for ω a scalar field. Now let ω_a be a vector field. Then

$$\begin{aligned}&\eta^m \mathscr{L}_\xi D(\omega)_{ma} - \eta^m D(\mathscr{L}_\xi \omega)_{ma} \\ &= \mathscr{L}_\xi(\eta^m D(\omega)_{ma}) - D(\omega)_{ma}\mathscr{L}_\xi \eta^m - \frac{1}{2}\mathscr{L}_\eta \mathscr{L}_\xi \omega_a + \frac{1}{2}D(\lambda)_a \\ &= \mathscr{L}_\xi\Big(\frac{1}{2}\mathscr{L}_\eta \omega_a - \frac{1}{2}D(\mu)_a\Big) - \Big(\frac{1}{2}\mathscr{L}_{(\mathscr{L}_\xi \eta)}\omega_a - \frac{1}{2}D(\nu)_a\Big) \\ &\quad - \frac{1}{2}\mathscr{L}_\eta \mathscr{L}_\xi \omega_a + \frac{1}{2}D(\lambda)_a \\ &= -\frac{1}{2}\mathscr{L}_\xi D(\mu)_a + \frac{1}{2}D(\nu)_a + \frac{1}{2}D(\lambda)_a \\ &= \frac{1}{2}D(-\mathscr{L}_\xi \mu + \nu + \lambda)_a = 0\end{aligned}$$

where we have set $\lambda = \eta^m \mathscr{L}_\xi \omega_m, \mu = \eta^a \omega_a$, and $\nu = \omega_a \mathscr{L}_\xi \eta^a$. Since η^a is arbitrary, (42) is satisfied for a 1-form. Therefore, for ω_a,

$$\begin{aligned}\xi^m D(D(\omega))_{mna} &= \frac{1}{3}\mathscr{L}_\xi D(\omega)_{na} - \frac{2}{3}D(\lambda)_{na} \\ &= \frac{1}{3}D(\mathscr{L}_\xi \omega)_{na} - \frac{2}{3}D(\lambda)_{na} = \frac{1}{3}D(\kappa)_{na}\end{aligned}$$

where $\lambda_a = \xi^m D(\omega)_{ma}$, and

$$\kappa_a = \mathscr{L}_\xi \omega_a - 2\xi^m D(\omega)_{ma} = D(\xi^m \omega_m)_a$$

Therefore, (43) is satisfied for ω_a a 1-form. Continuing in this way, proving first (42) and then (43) for each rank, we see that (42) and (43) are true in general.

One further property of the exterior derivative remains to be shown. Eqn. (38) suggests that the exterior derivative satisfies a "Leibnitz rule" under outer product. However, the outer product of two forms, e.g., $\alpha_{a...c}\beta_{d...f}$, is not in general antisymmetric. (In fact, it is antisymmetric if and only if one factor is 0–form or one factor is zero.) To correct this, we take the outer product and antisymmetrize, i.e., $\alpha_{[a...c}\beta_{d...f]}$ This is called the wedge product of α and β. We now prove the "generalized Leibnitz rule" for exterior derivatives:

$$D(\omega)_{ma...f} = D(\alpha)_{[ma...c}\beta_{d...f]} + D(\beta)_{[md...f}\alpha_{a...c]} \tag{44}$$

where $\omega_{a...f} = \alpha_{[a...c}\beta_{d...f]}$. Let p and q be the ranks of $\alpha_{a...c}$ and $\beta_{d...f}$, respectively, and set

$$\mu_{b...f} = \xi^a\alpha_{ab...c} \qquad \nu_{c...f} = \xi^d\beta_{dc...f}$$

$$\lambda_{b...f} = \xi^a\alpha_{a...f} = \frac{p}{p+q}\mu_{[b...c}\beta_{d...f]} + \frac{(-1)^{p-1}q}{p+q}\alpha_{[b...cd}\nu_{e...f]}$$

Then

$$\begin{aligned}
\xi^m D(\omega)_{ma...f} &= \frac{1}{p+q+1}\mathscr{L}_\xi\omega_{a...f} - \frac{p+q}{p+q+1}(D)\lambda_{ab...f}\\
&= \frac{1}{p+q+1}\alpha_{[a...c}\mathscr{L}_\xi\beta_{d...f]} + \frac{1}{p+q+1}\beta_{[d...f}\mathscr{L}_\xi\alpha_{a...c]} - \frac{p}{p+q+1}D(\nu)_{[a...c}\beta_{d...f]}\\
&\quad - \frac{p}{p+q+1}D(\beta)_{[ad...f}\mu_{b...c]} - \frac{(-1)^{p-1}q}{p+q+1}D(\alpha)_{[ab...d}\nu_{e...f]} - \frac{(-1)^{p-1}q}{p+q+1}D(\nu)_{[ac...f}\alpha_{b...d]}\\
&= \xi^m\Big(D(\alpha)_{[ma...c}\beta_{d...f]} + D(\beta)_{[md...f}\alpha_{a...c]}\Big)
\end{aligned}$$

which proves (44)

We summarize:

Theorem 9. The exterior derivative satisfies

ED1. $D(\omega+\mu)_{ma...d} = D(\omega)_{ma...d} + D(\mu)_{ma...d}$

ED2. $D(\omega)_{ma...d} = D(\alpha)_{[ma...b}\beta_{c...d]} + D(\beta)_{[mc...d}\alpha_{a...b]}(\omega_{a...d} = \alpha_{[a...b}\beta_{c...d]})$

ED3. $D(D(\omega))_{mna...d} = 0$

Furthermore, the exterior derivative and Lie derivative are related by

ELD1. $\mathscr{L}_\xi D(\omega)_{ma...d} = D(\mathscr{L}_\xi\omega)_{ma...d}$

ELD2. $\mathscr{L}_\xi\omega_{a...d} - pD(\lambda)_{a...d} - (p+1)\xi^m D(\omega)_{ma...d} = 0$

where, in ELD2, we have set $\lambda_{b...d} = \xi^a\omega_{ab...d}$, and where $\omega_{a...d}$ is a p–form.

Exercise 39. Write the exterior derivative in terms of components.

Example 27. In electrodynamics, the vector potential is a 1-form, A_b. The electromagnetic field is given by $F_{ab} = D(A)_{ab}$. It follows from ED3 that $D(F)_{mab} = 0$. This is one of Maxwell's equations.

Example 28. In classical mechanics, the symplectic structure on phase space is given by a 2-form, S_{ab}, with $D(S)_{mab} = 0$. A vector field ξ^a on phase space space is said to generate an infinitesimal canonical transformation if $\mathcal{L}_\xi S_{ab} = 0$. Setting $\lambda_b = \xi^a S_{ab}$, we see from ELD2 that, if ξ^a generates an infinitesimal canonical transformation, then $D(\lambda)_{mb} = 0$. The Hamiltonian H is a function on phase space. Choose ξ^a such that $\xi^a S_{ab} = D(H)_b$. Then, from ED3, ξ^a automatically generates an infinitesimal canonical transformation. This transformation describes the dynamics of the system.

10. Derivative Operators

We have now seen two contexts in which derivatives are applicable to tensor fields: Lie derivatives and exterior derivatives. The first of these requires a vector field ξ^a; the second is applicable only to tensor fields with a certain index structure. The time has now come to ask about derivative-type operations which satisfy all the nice properties one would ask of such an operation.

By a *derivative operator*, we understand a mapping, ∇_a, from tensor fields to tensor fields, where $\nabla_a \alpha_{m\ldots}{}^{c}$ has one more covariant index than $\alpha_{m\ldots}{}^{c}$, this mapping subject to the following conditions:

DO1. $\nabla_a(\alpha_{m\ldots}{}^{c} + \beta_{m\ldots}{}^{c}) = \nabla_a \alpha_{m\ldots}{}^{c} + \nabla_a \beta_{m\ldots}{}^{c}$.

DO2. $\nabla_a(\alpha_{m\ldots}{}^{c} \mu^{n\ldots}{}_{d}) = \alpha_{m\ldots}{}^{c} \nabla_a \mu^{n\ldots}{}_{d} + \mu^{n\ldots}{}_{d} \nabla_a \alpha_{m\ldots}{}^{c}$.

DO3. The derivative operator ∇_a can applied before or after contraction, with the same result. The derivative operator ∇_a can be applied before or after index substitution, with the same result.

DO4. For any scalar field α, $\nabla_a \alpha = D(\alpha)_a$.

DO5. For any scalar field α, $\nabla_{[a} \nabla_{b]} \alpha = 0$.

Of course, derivative operators ∇_b, ∇_c, etc. are induced from ∇_a by index substitution. The operator ∇_a cannot be applied to a tensor field having "a" as a covariant index, i.e., we cannot write $\nabla_a \alpha_m{}^{bc}{}_{ap}$. If ∇_a is applied to a tensor field with "a" as a contravariant index, then contraction is understood. That is, if $\beta_{am}{}^{bc}{}_{p} = \nabla_a \alpha_m{}^{bc}{}_{p}$. Then we write $\nabla_a \alpha_m{}^{ba}{}_{p}$ for $\beta_{am}{}^{ba}{}_{p}$. Condition DO1 is, of course, natural for a derivative operator, while DO2 is the Leibnitz rule again. Conditions DO1 and DO2 are characteristic of things which "behave like derivatives". In more detail, the first statement in condition DO3 means the following. If $\beta_{am}{}^{bc}{}_{p} = \nabla_a \alpha_m{}^{bc}{}_{p}$, then $\beta_{am}{}^{bm}{}_{p} = \nabla_a(\alpha_m{}^{bm}{}_{p})$. There is only one reasonable type of "derivative" of a scalar field, namely, the gradient. Condition DO4 ensures that the derivative operator, applied to a scalar field, is just this gradient. Condition DO5 states that two derivative operators can be applied to a scalar field in either order (i.e., $\nabla_a \nabla_b \alpha = \nabla_b \nabla_a \alpha$). That is to say, derivatives commute on scalar fields. One could require DO5 for all tensor fields, but this turns out to eliminate most interesting and useful cases. Occasionally, one sees the notion of a derivative operator introduced with condition DO5 omitted. These "generalized derivative operators" are said to have torsion, and, when DO5 happened to be satisfied, the operator is said to be torsion-free. We shall see shortly that "derivative operators with torsion" are easily treated within the context of the definition above.

Example 29. Consider the n−dimensional manifold R^n, with its natural chart (Example 1). Then a tensor field is uniquely characterized by its components with respect to this chart. If, for example, $\alpha_a{}^{bc}$ is a tensor field, and $\alpha_i{}^{jk}(x)$ $(i, j, k = 1, 2, \ldots, n)$ are its components, then denote by $\nabla_d \alpha_a{}^{bc}$ the tensor field whose components, with respect to this chart, are $\frac{\partial}{\partial x^m}\alpha_i{}^{jk}(x)$ $(i, j, k, m = 1, 2, \ldots, n)$. This ∇_d is a derivative operator on R^n, i.e., it satisfies DO1–DO5.

The natural question to ask, in response to the definition above, are those concerning the existence and uniqueness of derivative operators on a given manifold M. These questions have complete, and relatively simple, answers. We begin with uniqueness.

Let ∇_a and ∇'_a be two derivative operators. Then, for any covariant vector fields μ_b and ν_b, and any scalar field α, we have

$$\begin{aligned}(\nabla'_a - \nabla_a)(\alpha\mu_b + \nu_b) &= \mu_b(\nabla'_a - \nabla_a)\alpha + \alpha(\nabla'_a - \nabla_a)\nu_b \\ &= \alpha(\nabla'_a - \nabla_a)\mu_b + (\nabla'_a - \nabla_a)\nu_b\end{aligned}$$

where, in the first step, we have used DO2, and in the second, DO4. Thus, the left side of

$$(\nabla'_a - \nabla_a)\mu_b - \gamma^m{}_{ab}\mu_m \tag{45}$$

is linear in μ_b. Therefore, there exists a tensor field $\gamma^m{}_{ab}$ such that Eqn. (45) holds for all μ_b. In particular, setting $\mu_b = \nabla_b f = \nabla'_b f$ in (45), and antisymmetrising over "a" and "b", using DO5), we have $\gamma^m{}_{[ab]}\nabla_m f = 0$ for all f. Hence, the tensor field $\gamma^m{}_{ab}$ necessarily satisfies

$$\gamma^m{}_{[ab]} = 0, \quad \text{i..e.,} \gamma^m{}_{ab} = \gamma^m{}_{(ab)} \tag{46}$$

For arbitrary ξ^b and μ_b, we have

$$\begin{aligned}0 = (\nabla'_a - \nabla_a)(\xi^b\mu_b) &= \xi^b(\nabla'_a - \nabla_a)\mu_b + \mu_b(\nabla'_a - \nabla_a)\xi^b \\ = \xi^b\gamma^m{}_{ab}\mu_m + \mu_m(\nabla'_a - \nabla_a)\xi^m\end{aligned}$$

where, for the first equality, we have used DO4, for the second, DO2 and DO3, and, for third (45). Since this equation holds for all μ_m, we have

$$(\nabla'_a - \nabla_a)\xi^m = -\gamma^m{}_{ab}\xi^b \tag{47}$$

for all \$xi^b. Finally, for a general field, e,g., $\alpha_b{}^{cd}$,

$$\begin{aligned}0 = (\nabla'_a - \nabla_a)(\alpha_b^{cd}\xi^b\mu_c\nu_d) &= \xi^b\mu_c\nu_d(\nabla'_a - \nabla_a)\alpha_b{}^{cd} \\ + \alpha_b{}^{cd}\mu_c\nu_d(\nabla'_a - \nabla_a)\xi^b &+ \alpha_b{}^{cd}\xi^b\nu_d(\nabla'_a - \nabla_a)\mu_c \\ + \alpha_b{}^{cd}\xi^b\mu_c(\nabla'_a - \nabla_a)\nu_d\end{aligned}$$

Using (45), (47), and the fact that ξ^b, μ_c, and ν_d are arbitrary, we obtain

$$\nabla'_a \alpha_b{}^{cd} = \nabla_a \alpha_b{}^{cd} + \gamma^m{}_{ab} \alpha_m{}^{cd} - \gamma^c{}_{am} \alpha_b{}^{md} - \gamma^d{}_{am} \alpha_b{}^{cm} \tag{48}$$

A similar formula holds, of course, for tensor fields of arbitrary rank.

We summarize:

Theorem 10. Let ∇'_a and ∇_a be derivative operators on the manifold M. Then there exists a tensor field $\gamma^m{}_{ab} = \gamma^m{}_{(ab)}$ such that Eqn. (48) holds for any tensor field, $\alpha_{b\ldots}{}^{d}$. Conversely, if ∇_a is any derivative operator on M, and $\gamma^m{}_{ab} = \gamma^m{}_{(ab)}$ is a tensor field on M, then ∇'_a, defined by (48), is also a derivative operator.

The first statement of Theorem 10 was proved above. The second is easy to verify: one has to check that ∇'_a, defined by (48), satisfies DO1 - DO5. [Exercise 40. Carry out this verification.] [Remark: The situation with regard to torsion should now be clear. The only place we used DO5 was to obtain (46). Thus, there is a one-to-one correspondence between operators which satisfy DO1-DO4 and pairs $(\nabla_a, \delta^m{}_{ab})$, where $\delta^m{}_{ab} = \delta^m{}_{[ab]}$, and ∇_a is a derivative operator. The dropping of DO5 merely permits one to "hide" an extra tensor field $\delta^m{}_{ab} = \delta^m{}_{[ab]}$ in his derivative operators.]

Theorem 10 is the whole story concerning uniqueness of derivative operators. A derivative operator is never unique (except when M is zero-dimensional): one can choose any old tensor field $\gamma^m{}_{ab} = \gamma^m{}_{(ab)} \neq 0$ and get another one. If one knows one derivative operator on a manifold, he knows them all. [underlineExercise 41. Write down the most general derivative operator on R^n.] The collection of all derivative operators on a manifold practically have the structure of a vector space, except that there is no natural "origin". (That is, there is no natural "zero derivative operator". If one randomly chooses some derivative operator, ∇_a, then, by Theorem 10, there is one-to-one correspondence between the collection of derivative operators on M and the vector space of tensor fields $\gamma^m{}_{ab} = \gamma^m{}_{(ab)}$.) The collection of all derivative operators on a manifold are a beautiful example of what is called an affine space.

We next consider existence of derivative operator on a manifold. The situation can be summarized by the following:

Theorem 11. A necessary and sufficient condition that a manifold M possess a derivative operator is the following: there exists a sub-collection of the collection of all the charts on M such that i) every point of M is in at least one of the U's in this sub-collection, and ii) no point of M is in an infinite number of U's in this sub-collection.

A manifold satisfying the conditions of Theorem 11 is said to be *paracompact.* [Paracompactness is actually a topological property which, in order to avoid topological questions, we have reformulated in terms of chart.] Every manifold one is ever likely to see is paracompact. In fact, it is a rather subtle business to even contract a manifold which is not paracompact. It is immediate from the definitions that the manifolds R^n and S^n are paracompact. [Exercise 42. Prove that the product of two paracompact manifolds paracompact. Exercise 43. Prove that the result of "cutting out" a closed region from a paracompact manifold is a paracompact manifold.] Although the proof of Theorem 11 is not very difficult, it is rather technical. Since neither the statement nor the proof of Theorem 11 is very important for differential geometry, we omit the proof.

Thus, every reasonable manifold has plenty of derivative operators, with any one easily obtainable from any other. A few pathological manifolds have no derivative operators.

Exercise 44. Introduce a chart. express the action of a derivative operator, ∇_a, in terms of an expression involving the components of the tensor field to which ∇_a is being applied. There will appear in your expression functions $\Gamma^i{}_{jk}(x)$, $i, j = 1.2, \ldots, n$. These functions are called the connection. Derive the (well-known) formula for the behavior of the connection under a coordinate transformation.

Exercise 45. Prove that every manifold has a derivative operator locally. More precisely, prove that, if p is any point of a manifold M, then there is an open subset O of M, containing p, such that the manifold $O = M - (M - O)$ possesses a derivative operator. (Thus, paracompactness is a "global property" of manifolds.)

Exercise 46. Why do not we introduce a "derivative operator" with a contravariant rather than a covariant index?

Exercise 47. Find a derivative operator on S^n.

Exercise 48. the derivative operator of Example 29 has the property that derivatives commute on an arbitrary tensor field. Verify that this property does not hold for every derivative operator on the manifold R^n.

Exercise 49 Let $\widetilde{\nabla}_a$ be an operator, defined only on covariant vector fields, which satisfies i) $\widetilde{\nabla}_a(\mu_b + \nu_b) - \widetilde{\nabla}_a\mu_b + \widetilde{\nabla}_a\nu_b$, ii) $\widetilde{\nabla}_a(\alpha\mu_b) = \alpha\widetilde{\nabla}_a\mu_b + \mu_b D(\alpha)_a$, iii) $\widetilde{\nabla}_{[a}D(\alpha)_{b]} = 0$. Prove that there exists precisely one derivative operator ∇_a which, on covariant vector fields, coincides with $\widetilde{\nabla}_a$.

11. Concomitants

Fix a derivative operator ∇_a and a vector field ξ^a. Then, from DO4, we have

$$\mathcal{L}_\xi \alpha = \xi^a \nabla_a \alpha \tag{49}$$

for every scalar field α. Now let η^a be another contravariant vector field. Then

$$\begin{aligned}
&\mathcal{L}_{(\mathcal{L}_\xi \eta)}\alpha = \mathcal{L}_\xi \mathcal{L}_\eta \alpha - \mathcal{L}_\eta \mathcal{L}_\xi \alpha = \xi^b \nabla_b(\eta^a \nabla_a \alpha) - \eta^b \nabla_b(\xi^a \nabla_a \alpha) \\
&= (\xi^b \nabla_b \eta^a)\nabla_a \alpha + \xi^b \eta^a \nabla_b \nabla_a \alpha - (\eta^b \nabla_b \xi^a)\nabla_a \alpha - \eta^b \xi^a \nabla_b \nabla_a \alpha \\
&= (\xi^b \nabla_b \eta^a - \eta^b \nabla_b \xi^a)\nabla_a \alpha
\end{aligned}$$

where we have used DO4 and DO5. Since α is arbitrary, we have

$$\mathcal{L}_\xi \eta^a = \xi^b \nabla_b \eta^a - \eta^b \nabla_b \xi^a \tag{50}$$

Now let μ_a be a covariant vector field. Then

$$\begin{aligned}
&\eta^a \mathcal{L}_\xi \mu_a = \mathcal{L}_\xi(\mu_a \eta^a) - \mu_a \mathcal{L}_\xi \eta^a = \xi^b \nabla_b(\mu_a \eta^a) \\
&- \mu_a(\xi^b \nabla_b \eta^a - \eta^b \nabla_b \xi^a) = \eta^a(\xi^b \nabla_b \mu_a + \mu_b \nabla_a \xi^b)
\end{aligned}$$

Since η^a is arbitrary, we have

$$\mathcal{L}_\xi \mu_a = \xi^b \nabla_b \mu_a + \mu_b \nabla_a \xi^b \tag{51}$$

Finally, consider an arbitrary tensor field, e.g., $\alpha^{ab}{}_{cd}$. Then

$$\begin{aligned}
&\mu_a \nu_b \eta^c \tau^d \mathcal{L}_\xi \alpha^{ab}{}_{cd} = \mathcal{L}_\xi(\alpha^{ab}{}_{cd}\mu_a \nu_b \eta^c \tau^d) - \alpha^{ab}{}_{cd}\nu_b \eta^c \tau^d \mathcal{L}_\xi \mu_a \\
&- \alpha^{ab}{}_{cd}\mu_a \eta^c \tau^d \mathcal{L}_\xi \nu_b - \alpha^{ab}{}_{cd}\mu_a \nu_b \tau^d \mathcal{L}_\xi \eta^c - \alpha^{ab}{}_{cd}\mu_a \nu_b \eta^c \mathcal{L}_\xi \tau^d
\end{aligned}$$

Substituting (50) and (51), and using the fact that μ_a, ν_b, $\eta^c$and τ^d are arbitrary, we have

$$\mathcal{L}_\xi \alpha^{ab}{}_{cd} = \xi^m \nabla_m \alpha^{ab}{}_{cd} - \alpha^{mb}{}_{cd}\nabla_m \xi^a - \alpha^{am}{}_{cd}\nabla_m \xi^b + \alpha^{ab}{}_{md}\nabla_c \xi^m + \alpha^{ab}{}_{cm}\nabla_d \xi^m \tag{52}$$

Thus, using an arbitrary derivative operator ∇_a, we have obtained an "explicit" expression for the Lie derivative.

It is clear from Eqn. (52) that the right side of this equation must be independent of the choice of the derivative operator ∇_a (for the left side certainly is). We verify this explicitly, using Eqn. (48), for the simpler case in which $\alpha^a{}_b$ has only two indices:

$$\begin{aligned}&\xi^m \nabla'_m \alpha^a{}_b - \alpha^m{}_b \nabla'_m \xi^a + \alpha^a{}_m \nabla'_b \xi^m = (\xi^m \nabla_m \alpha^a{}_b - \xi^m \gamma^a{}_{mn} \alpha^n{}_b \\ &+ \xi^m \gamma^n{}_{mb} \alpha^a{}_n) - (\alpha^m{}_b \nabla_m \xi^a - \alpha^m{}_b \gamma^a{}_{mn} \xi^n) + (\alpha^a{}_m \nabla_b \xi^m - \alpha^a{}_m \gamma^m{}_{bn} \xi^n) \\ &= \xi^m \nabla_m \alpha^a{}_b - \alpha^m{}_b \nabla_m \xi^a + \alpha^a{}_m \nabla_b \xi^m\end{aligned}$$

An expression involving some tensor fields and a derivative operator ∇_a, which is independent of the choice of derivative operator, is called a *concomitant*. Thus, the right side of (52) is a concomitant.

Let $\omega_{a...c}$ be antisymmetric in all indices. We show that the expression

$$\nabla_{[m}\omega_{a...c]} \tag{53}$$

is a concomitant. We first use (48):

$$\nabla'_m \omega_{a...c} = \nabla_m \omega_{a...c} + \gamma^n{}_{ma} \omega_{n...c} + \cdots + \gamma^n{}_{mc} \omega_{a...n}$$

Antisymmetrizing over "$ma \dots c$", and using (46), the result follows immediately. We now claim that the concomitant (53) is precisely the exterior derivative, i.e., that

$$D(\omega)_{ma...c} = \nabla_{[m}\omega_{a...c]} \tag{54}$$

For ω_a a scalar field, this is clear. For ω_a a vector field, set $\lambda = \omega_a \xi^a$. Then

$$\begin{aligned}\xi^m D(\omega)_{ma} &= \frac{1}{2}\mathcal{L}_\xi \omega_a - \frac{1}{2}D(\lambda)_a = \frac{1}{2}(\xi^m \nabla_m \omega_a + \omega_m \nabla_a \xi^m) \\ &- \frac{1}{2}\nabla_a(\xi^m \omega_m) = \frac{1}{2}\xi^m \nabla_m \omega_a - \frac{1}{2}\xi^m \nabla_a \omega_m = \xi^m \nabla_{[m}\omega_{a]}\end{aligned}$$

Similarly for higher rank.

In Sections 7 and 9, we defined the Lie derivative and exterior derivative "algebraically". We now have expressions for these things as concomitants. We could just as well have defined the Lie derivative by (52) and the exterior derivative by (54) (after observing that the right sides of these expressions are concomitants). The various properties of the Lie and exterior derivatives are, as a rule, somewhat easier to derive from (52) and (54) than from the definitions in Sects. 7 and 9. Properties LD1, LD2, LD3, LD4, and Eqn. (29) follow from Eqn. (52) by inspection. Eqn. (31) for Lie derivatives follows from (52) after a few lines of calculation. (Exercise 50. Check this.) Properties ED1 and ED2 follow from (54) by inspection, while ELD1 and

ELD2 require a few lines of calculation, and ED3 is a bit tricker (although ED3, too, will become simple after the following section). (Exercise 51. Verify ELD1 and ELD2 from (52) and (54).)

Exercise 52. Using the remarks above, together with Example 29, write down the answers to Exercises 34 and 39. (No calculations whatever are required.)

Concomitants often provide a "brute-force" way of doing things.

Example 30. We give another example of a concomitant. Let $\alpha^{a_1\ldots a_p}$ and $\beta^{a_1\ldots a_q}$ be symmetric in all p indices and all q indices, respectively. Set

$$[\alpha,\beta] = p\,\alpha^{m(a_1\ldots a_{p-1}}\nabla_m\beta^{a_p\ldots a_{p+q-1})} - q\beta^{m(a_1\ldots a_{q-1}}\nabla_m\alpha^{a_q\ldots a_{p+q-1})} \tag{55}$$

(Why this notation? What is (55) when $p = q = 1$?) The right side of (55) is a concomitant:

$$\begin{aligned}
&p\,\alpha^{ma_1\ldots a_{p-1}}\nabla'_m\beta^{a_p\ldots a_{p+q-1}} - q\beta^{ma_1\ldots a_{q-1}}\nabla'_m\alpha^{a_q\ldots a_{p+q-1}}\\
&= p\,\alpha^{ma_1\ldots a_{p-1}}(\nabla_m\beta^{a_p\ldots a_{p+q-1}} - \gamma^{ap}_{mn}\beta^{n\ldots a_{p+q-1}} - \cdots - \gamma^{a_{p+q-1}}_{mn}\beta^{a_p\ldots n})\\
&- q\beta^{ma_1\ldots a_{q-1}}(\nabla_m\alpha^{a_q\ldots a_{p+q-1}} - \gamma^{a_q}_{mn}\alpha^{n\ldots a_{p+q-1}} - \cdots - \gamma^{a_{p+q-1}}_{mn}\alpha^{a_q\ldots n})
\end{aligned}$$

Symmetrizing over the indices "$a_1 \ldots a_{p+q-1}$", we see that all the γ^m_{ab}–terms, cancel out. (A total of q terms arise from $\nabla_m\beta^{\cdots}$, and p terms from $\nabla_m\alpha^{\cdots}$. All these terms are equal, so they all cancel.) This concomitant has some pretty properties. In fact it satisfies (24) and (25). (Thus, we have a Lie algebra of totally symmetric contravariant tensor fields.) There is also a "Leibnitz rule". We define the only kind of "outer product" we can with these tensor:

$$\alpha\cap\beta = \alpha^{(a_1\ldots a_p}\beta^{a_{p+1}\ldots a_{p+q+1})} \tag{56}$$

Then

$$(\alpha,\beta\cap\gamma) = (\alpha,\beta)\cap\gamma + (\alpha,\gamma)\cap\beta \tag{57}$$

(Exercise 53. Verify that the concomitant (55) satisfies (24), (25), and (57).)

Exercise 54. Let $\alpha^{a_1\ldots a_p}$ and $\beta^{a_1\ldots a_q}$ be antisymmetric in all p indices and all q indices, respectively. Prove that the expression

$$p(-1)^{p+1}\alpha^{m[a_1\ldots a_{p-1}}\nabla_m\beta^{a_p\ldots a_{p+q-1}]} - q(-1)^{p+pq}\beta^{m[a_1\ldots a_{q-1}} - \nabla_m\alpha^{a_q\ldots a_{p+q-1}]} \tag{58}$$

is a concomitant.

Exercise 55. Try to make a natural-looking algebra out of the concomitant of Exercise 54, in the same way as was done in Example 30.

Exercise 56. Let $\omega_{a\ldots c}$ be a tensor field (not necessarily antisymmetric) Is $\nabla_{[m}\omega_{a\ldots c]}$ necessarily a concomitant?

Although by far the most important concomitants are the Lie derivative and the exterior derivative, the other two described above are also useful for certain applications. Although many concomitants have been written down, there is, as far as I am aware, no systematic procedure known for finding them all.

12. Curvature

The natural thing to do with derivatives is commute them. By commuting derivative operators, we are led to curvature.

Let ∇_a be a derivative operator. Then

$$\nabla_{[a}\nabla_{b]}\left[\alpha\mu_c + \nu_c\right] = \nabla_{[a}\left[\alpha\nabla_{b]}\mu_c + (\nabla_{b]}\alpha)\mu_c + \nabla_{b]}\nu_c\right] = (\nabla_{a[}\alpha)\nabla_{b]}\mu_c$$
$$+ \alpha\nabla_{[a}\nabla_{b]}\mu_c + \mu_c\nabla_{[a}\nabla_{b]}\alpha + (\nabla_{[b}\alpha)\nabla_{a]}\mu_c + \nabla_{[a}\nabla_{b]}\nu_c$$
$$= \alpha\nabla_{[a}\nabla_{b]}\mu_c + \nabla_{[a}\nabla_{b]}\nu_c$$

that is to say, the left side of

$$\nabla_{[a}\nabla_{b]}\mu_c = \frac{1}{2}R_{abc}{}^{d}\mu_d \tag{59}$$

is linear in μ_c. Hence, there exists a tensor field, $R_{abc}{}^{d}$, such that (59) holds for every μ_c. Furthermore,

$$0 = \nabla_{[a}\nabla_{b]}\left[\xi^c\mu_c\right] = \nabla_{[a}\left[(\nabla_{b]}\xi^c)\mu_c + (\nabla_{b]}\mu_c)\xi^c\right] = \mu_c\nabla_{[a}\nabla_{b]}\xi^c$$
$$+ (\nabla_{[b}\xi^c)\nabla_{a]}\mu_c + \xi^c\nabla_{[a}\nabla_{b]}\mu_c + (\nabla_{[a}\xi^c)\nabla_{b]}\mu^c$$
$$= \mu_c\nabla_{[a}\nabla_{b]}\xi^c + \xi^c\nabla_{[a}\nabla_{b]}\mu_c$$

Substituting (59) into the last expression, and using the fact that μ_c is arbitrary.

$$\nabla_{[a}\nabla_{b]}\xi^c = -\frac{1}{2}R_{abd}{}^{c}\xi^d \tag{60}$$

for every ξ^c. By the standard argument, we have, for a general tensor field, e.g., $\alpha^{cd}{}_{ps}$,

$$\nabla_{[a}\nabla_{b]}\alpha^{cd}{}_{ps} = -\frac{1}{2}R_{abm}{}^{c}\alpha^{md}{}_{ps} - \frac{1}{2}R_{abm}{}^{d}\alpha^{cm}{}_{ps} + \frac{1}{2}R_{abp}{}^{m}\alpha^{cd}{}_{ms} + \frac{1}{2}R_{abc}{}^{m}\alpha^{cd}{}_{pm} \tag{61}$$

The tensor field $R_{abc}{}^{d}$ is called the *Riemann tensor* (or curvature tensor) associated with the derivative operator ∇_a. It is immediate from the definition, e.g., (59) that

$$R_{abc}{}^{d} = R_{[ab]c}{}^{d} \tag{62}$$

There is one further algebraic condition on the Riemann tensor. Antisymmetrizing (59) over "a, b, c", we have $\nabla_{[a}\nabla_b\mu_{c]} = \frac{1}{2}R_{[abc]}{}^d\mu_d$ But Eqn. (54) and ED3 imply that the left side of this equation vanishes. Hence, since μ_c is arbitrary,

$$R_{[abc]}{}^d = 0 \tag{63}$$

Eqn. (62) and (63) are the only algebraic conditions on the Riemann tensor of a general derivative operator. There is, however, an additional differential condition. Using (59),

$$\begin{aligned}\nabla_a\nabla_{[b}\nabla_{c]}\mu_d &= \frac{1}{2}\nabla_a(R_{bcd}{}^m\mu_m)\\ &= \frac{1}{2}(\nabla_a R_{bcd}{}^m)\mu_m + \frac{1}{2}R_{bcd}{}^m\nabla_a\mu_m\end{aligned}$$

Using (61),

$$\nabla_{[a}\nabla_{b]}\nabla_c\mu_d = \frac{1}{2}R_{abc}{}^m\nabla_m\mu_d + \frac{1}{2}R_{abd}{}^m\nabla_c\mu_m$$

Now antisymmetrize these two equations over "a, b, c". The left sides become equal; equate the right sides. Then two terms cancel, while another is eliminated by (63). Thus, we are left with $\frac{1}{2}(\nabla_{[a}R_{bc]d}{}^m)\mu_m = 0$. Since μ_m is arbitrary,

$$\nabla_{[a}R_{bc]d}{}^m = 0 \tag{64}$$

Eqn. (64) is called the *Bianchi identity*.

To summarize, given any derivative operator on a manifold, there exists a tensor field field $R_{abc}{}^d$ such that (61) holds. This Riemann tensor satisfies the algebraic equations (62) and (63), as well as the differential equation (64).

Exercise 57. Prove that Riemann tensor associated with the derivative operator of Example 29 is zero.

Example 31. Let ∇_a and ∇'_a be derivative operators, related via Eqn. (48), and let $R_{abc}{}^d$ and $R'_{abc}{}^d$ be their respective Riemann tensors. We find an expression for $R'_{abc}{}^d$ in terms of $R_{abc}{}^d$ and γ^m_{ab}. For arbitrary μ_c,

$$\begin{aligned}&\nabla'_a\nabla'_b\mu_c = \nabla'_a(\nabla_b\mu_c + \gamma^m{}_{bc}\mu_m) = \nabla_a(\nabla_b\mu_c + \gamma^m{}_{bc}\mu_m)\\ &+ \gamma^n{}_{ab}(\nabla_n\mu_c + \gamma^m{}_{nc}\mu_m) + \gamma^n{}_{ac}(\nabla_b\mu_n + \gamma^m{}_{bn}\mu_m)\\ &= \nabla_a\nabla_b\mu_c + (\nabla_a\gamma^m{}_{bc})\mu_m + \gamma^m{}_{bc}\nabla_a\mu_m\\ &+ \gamma^n{}_{ab}(\nabla_n\mu_c + \gamma^m{}_{nc}\mu_m) + \gamma^n{}_{ac}\nabla_b\mu_n + \gamma^n{}_{ac}\gamma^m{}_{bn}\mu_m\end{aligned}$$

Now antisymmetrize this equation over "a" and "b". The fourth term on the right vanishes by (46), while the third and fifth terms cancel. Hence,

$$\nabla'_{[a}\nabla'_{b]}\mu_c = \nabla_{[a}\nabla_{b]}\mu_c + \mu_m\nabla_{[a}\gamma_{b]}{}^m{}_c + \mu_m\gamma^n{}_{c[a}\gamma^m{}_{b]n}$$

Now use (59) and the fact that μ_m is arbitrary:

$$R'_{abc}{}^{d} = R_{abc}{}^{d} + 2\nabla_{[a}\gamma_{b]}{}^{d}{}_{c} + 2\gamma^{n}{}_{c[a}\gamma_{b]}{}^{d}{}_{n} \tag{65}$$

This is the desired result.

Exercise 58. Verify directly from the right side of (65) that $R'_{[abc]}{}^{d} = 0$.

Exercise 59. Verify directly that right side of (65) satisfies the Bianchi identity.

Exercise 60. Derive the standard textbook formula for the components of the Riemann tensors in terms of derivatives of the connection with respect to the coordinates. (Use Exercise 57 and Example 31.)

Since derivatives should be commuted, let's commute a Lie derivative and a derivative operator.

$$\begin{aligned}
&\mathcal{L}_\xi(\nabla_a\mu_b) - \nabla_a(\mathcal{L}_\xi\mu_b) = \xi^m\nabla_m\nabla_a\mu_b + (\nabla_m\mu_b)\nabla_a\xi^m + (\nabla_a\mu_m)\nabla_b\xi^m \\
&- \nabla_a(\xi^m\nabla_m\mu_b + \mu_m\nabla_b\xi^m) = \xi^m\nabla_m\nabla_a\mu_b + (\nabla_m\mu_b)\nabla_a\xi^m + (\nabla_a\mu_m)\nabla_b\xi^m \\
&- \xi^m\nabla_a\nabla_m\mu_b - (\nabla_m\mu_b)(\nabla_a\xi^m) - \mu_m\nabla_a\nabla_b\xi^m - (\nabla_a\mu_m)(\nabla_b\xi^m) \\
&= 2\xi^m\nabla_{[m}\nabla_{a]}\mu_b - \mu_m\nabla_a\nabla_b\xi^m = (R_{nab}{}^{m}\xi^n - \nabla_a\nabla_b\xi^m)\mu_m
\end{aligned}$$

Thus, $\mathcal{L}_\xi$ and ∇_a commute, on all μ_b, if and only if

$$-\nabla_a\nabla_b\xi^m + R_{nab}{}^{m}\xi^n = 0 \tag{66}$$

in analogy with (31), we may regard the left side of (66) as the "Lie derivative of ∇_a in the xi^a-direction". (A solution ξ^a of (66) is called an affine collineation.)

Exercise 61. Check that $\mathcal{L}_\xi$ and ∇_a commute when applied to an arbitrary tensor field if and only if (66) is satisfied.

Exercise 62. Let ξ^a be an affine collineation for a derivative operator ∇_a, and let ∇'_a be another derivative operator, related to ∇_a via (48). Prove that ξ^a is an affine collineation for ∇'_a if and only if $\mathcal{L}_\xi\gamma^m{}_{ab} = 0$. (this result further strengthens the remark above concerning the interpretation of the left side of (66).)

Exercise 63. Prove that $R_{abm}{}^{m} = -2R_{m[ab]}{}^{m}$. Using this result, check that the result of antisymmetrizing (66) over "a" and "b" is an identity. Hence, Eqn. (66) is equivalent to that equation symmetrized over "a, b". (In this form, the left side of (66) has the same structure as $\gamma^m{}_{ab}$.)

Exercise 64. Using (65), verify that there are no algebraic conditions on the Riemann tensor other than (62) and (63). That is, show that, given, given a tensor $R_{abc}{}^{d}$ at a point p, is satisfying (62) and (63), there exists a derivative operator whose Riemann tensor at p is $R_{abc}{}^{d}$.

13. Metrics

Let M be a manifold. A *metric* on M consists of a tensor field g_{ab} which is symmetric (i.e., $g_{ab} - g_{(ab)}$) and invertible (i.e., there exists a tensor field g^{ab} such that $g^{ac}g_{bc} = \delta^a{}_c$) The study of manifolds with metrics is called Riemann geometry.

We verify that the inverse, g^{ab}, of a metric g_{ab} is symmetric and unique. Symmetry follows from $g^{ac}g^{bd}g_{cd} = g^{ac}\delta^b{}_c = g^{ab}$ and the observation that, since g_{cd} is symmetric, so is the left side of this equation. To prove uniqueness, let g^{ab} and g'^{ab} be two inverses for g_{ab}. Then

$$g^{ab} = g^{ac}\delta^b{}_c = g^{ac}(g'^{bd}g_{cd}) = g'^{bd}\delta^a{}_d = g'^{ab}$$

Let M be a manifold with metricg_{ab}. It is convenient to incorporate this metric into the index notation as follows. If, e.g., $\alpha_{bs}{}^{rc}{}_d{}^v$ is any tensor field on M, we set

$$\begin{aligned}
\alpha_{bs}{}^{r}{}_{cd}{}^{v} &= g_{cm}\alpha_{bs}{}^{rm}{}_{d}{}^{v} \\
\alpha^{b}{}_{s}{}^{rc}{}_{d}{}^{v} &= g^{bm}\alpha_{ms}{}^{rc}{}_{d}{}^{v} \\
\alpha_{bs}{}^{rc}{}_{dv} &= g_{vm}\alpha_{bs}{}^{rc}{}_{d}{}^{m} \\
\alpha_{bs}{}^{rcdv} &= g^{dm}\alpha_{bs}{}^{rc}{}_{m}{}^{v}
\end{aligned}$$

This operation is called *raising and lowering of indices*. Evidently, the result of first raising an index and then lowering that index is to leave the tensor invariant. Note that the result of raising or lowering an index of a tensor field depends on the choice of metric. [If two metrics appear on a manifold, one must either suspend this convention, or select one of the metrics to be the one to be used in raising and lowering.] The use of this convention has the consequence that the metric will almost never appear explicitly.

Let g_{ab} be a metric on M and p on a point of M. The metricg_{ab} is said to have *signature* (n^+, n^-) [n^+ and n^- are non-negative integers with $n^+ + n^- = n$, the dimension of M] at p if there are n^+ vectors $\underset{1}{\xi}^a, \ldots, \underset{n^+}{\xi}^a$ at p, and n^-

vectors $\underset{1}{\eta}^a, \ldots, \underset{n^-}{\eta}^a$ at p such that

$$\underset{i}{\xi}^a \underset{j}{\eta}_a = 0 \quad (i = 1, \ldots, n^+;\ j = 1, \ldots, n^-)$$

$$\underset{i}{\xi}^a \underset{j}{\xi}_a = 0 \quad (i, j = 1, \ldots, n^+;\ i \neq j)$$

$$\underset{i}{\eta}^a \underset{j}{\eta}_a = 0 \quad (i, j = 1, \ldots, n^-;\ i \neq j)$$

$$\underset{i}{\xi}^a \underset{j}{\xi}_a = +1 \quad (i = j = 1, \ldots, n^+)$$

$$\underset{i}{\eta}^a \underset{j}{\eta}_a = -1 \quad (i = j = 1, \ldots, n^-)$$

Introducing a chart, the components of g_{ab} at p form an $n \times n$ matrix, which is invertible. Since every such matrix can be diagonalized, with diagonal elements all either $+1$ or -1, and since the number of $+1$'s and -1's is independent of the diagonalization, it is clear that a metric g_{ab} has a unique signature at each point p. (Of course, the ξ's and η's are not unique.) If M is n-dimensional, there are $(n+1)$ possible signature, $(0, n), (1, n-1), \ldots, (n, 0)$. Since, in terms of a chart, the components of g_{ab} are continuous, and since the matrix of components is invertible at each point, it follows that, if g_{ab} has signature $(n^+,\ n^-)$ at p, there is an open subset O of M containing p such that g_{ab} has signature $(n^+,\ n^-)$ at each point of O. Thus, the set of points of M at which g_{ab} has any given signature is open. Since the set of points at which g_{ab} has any other signature is also open, the set of points at which g_{ab} has any given signature is also closed. A manifold M is said to be *connected* if the only subsets of M which are simultaneously open and closed are the empty set and M itself. Thus, a metric on a connected manifold M has the same signature at every point. (Exercise 65. Verify that R^n is connected, and that S^n is connected for $n \geq 1$. Exercise 66. Prove that the product of two connected manifolds M is connected. Exercise 67. Find an example of a connected manifold M and closed subset C such that $M - C$ is not connected.)

A metric g_{ab} on M is said to be *positive-definite* if g_{ab} has signature $(n,\ 0)$, *negative-definite* if signature $(0,\ n)$, and *indefinite* if some other signature, where, in each case, these terms are used only if g_{ab} has the same signature at all points of M. The term "Riemannian geometry" is sometimes used only when g_{ab} is positive-definite, and "pseudo-Riemannian geometry" otherwise.

Exercise 68. Let the metric g_{ab} on M have signature $(n^+,\ n^-)$ (everywhere). Prove that $-g_{ab}$ is also a metric on M, and that this metric has signature $(n^-,\ n^+)$.

Exercise 69. Let g_{ab} be a positive-definite metric on M, p a point of M. Prove that, for every vector ξ^a at p, $\xi^a \xi_a \geq 0$, with equality holding if and only if $\xi^a = 0$.

Exercise 70. Let g_{ab} be an indefinite metric on M, p a point of M.

Prove that there exists a nonzero vector ξ^a at p that $\xi^a \xi_a = 0$.

Exercise 71. State and prove the triangle and Schwarz inequalities for vectors at a point of a manifold with positive-definite metric.

Example 32. Let g_{ab} and g'_{ab} be metric on M, and suppose that, for every ξ^a, $g_{ab}\xi^a\xi^b = g'_{ab}\xi^a\xi^b$. We show that $g_{ab} = g'_{ab}$. Replacing ξ^a above by $\xi^a + \eta^a$, we have

$$g_{ab}\xi^a\xi^b + 2g_{ab}\xi^a\eta^b + g_{ab}\eta^a\eta^b = g'_{ab}\xi^a\xi^b + 2g'_{ab}\xi^a\eta^b + g'_{ab}\eta^a\eta^b$$

But the first term on the left equals the first term on the right, and the last term on the left equals the last term on the right. Hence, $(g_{ab} - g'_{ab})\xi^a\eta^b = 0$ for all ξ^a, η^b. Hence, $g_{ab} = g'_{ab}$.

Example 33. We construct a positive-definite metric on S^n $(n > 0)$. Let U be the region $y_{n+1} > 0$, and let the coordinates be $x^1 = y_1, \ldots, x^n = y_n$. If ξ^a is a vector field on S^n, let $\xi^1(x), \ldots, \xi^n(x)$ be its components with respect to this chart, and let g_{ab} be such that $g_{ab}\xi^a\xi^b$ becomes, in this chart, the function $[(\xi')^2 + \ldots + (\xi^n)^2][1 - (x^1)^2 - \ldots - (x^n)^2]^{-1}$. Exercise 72. Prove that this defines a metric on S^n.

Exercise 73. Find a positive-definite metric on R^n, an indefinite metric on R^n.

One of the most important facts in Riemannian geometry is the following:

Theorem 12. Let M be a manifold with metric g_{ab}. Then there exists precisely one derivative operator ∇_a on M such that $\nabla_a g_{bc} = 0$.

Before proving Theorem 12, we give an argument which leads to the desired conclusion under the additional assumption that there exists at least one derivative operator, ∇'_a via (48). Then

$$\nabla'_a g_{bc} = \nabla_a g_{bc} + \gamma^m{}_{ab} g_{mc} + \gamma^m{}_{ac} g_{bm}$$

We wish to choose ∇_a, i.e., choose $\gamma^m{}_{ab}$, such that $\nabla_a g_{bc} = 0$. Thus, we have only to prove that there exists precisely one tensor field $\gamma^m{}_{ab} = \gamma^m{}_{(ab)}$ on M such that

$$\nabla'_a g_{bc} = \gamma^m{}_{ab} g_{mc} + \gamma^m{}_{ac} g_{bm} \quad (= 2\gamma_{(bc)a}) \tag{67}$$

There certainly exists such a tensor field, namely

$$\gamma^m{}_{ab} = \frac{1}{2} g^{mn}(\nabla'_a g_{bn} + \nabla'_b g_{an} - \nabla'_n g_{ab}) \tag{68}$$

(Exercise 74. Substitute (68) into (67).) To see that this solution is unique, let $\tilde{\gamma}^m{}_{ab}$ be another solution of (67), and set $\mu^m{}_{ab} = \tilde{\gamma}^m{}_{ab} - \gamma^m{}_{ab}$. Then

${\mu^m}_{ab} = {\mu^m}_{(ab)}$, and from (67), $\mu_{(ma)b} = 0$. To prove that ${\mu^m}_{ab} = 0$, we "interchange indices of μ_{mab}", first the first pair, then the second, etc.:

$$\begin{aligned} \mu_{mab} &= -\mu_{amb} = -\mu_{abm} = \mu_{bam} = \mu_{bma} \\ &= -\mu_{mba} = -\mu_{mab} \end{aligned}$$

This argument is interesting becausc it bring out clearly why Theorem 12 is true, namely, because ${\gamma^m}_{ab} = {\gamma^m}_{(ab)}$ and $\nabla'_a g_{bc}$ have the same index structure.

Proof of Theorem 12: For a scalar field α, set $\nabla_a \alpha = D(\alpha)_a$. For a covariant vector field, μ_b, set

$$\nabla_a \mu_b = \frac{1}{2} \mathscr{L}_\mu g_{ab} + D(\mu)_{ab}$$

where $\mu^b = g^{ab}\mu_a$. By (29) and (41), this satisfies $\nabla_a(\alpha\mu_b) = \alpha\nabla_a\mu_b + \mu_b\nabla_a\alpha$. For a contravariant vector field ξ^a, set

$$\nabla_a \xi^b = g^{bm} \nabla_a (g_{mn} \xi^n)$$

Finally, for an arbitrary tensor field, e.g., ${\alpha^{cd}}_{pq}$, define its derivative by

$$\begin{aligned} \mu_c \nu_d \xi^p \eta^q \nabla_a {\alpha^{cd}}_{pq} &= \nabla_a ({\alpha^{cd}}_{pq} \mu_c \nu_d \xi^p \eta^a) - {\alpha^{cd}}_{pq} \nu_d \xi^p \eta^q \nabla_a \mu_c \\ &- {\alpha^{cd}}_{pq} \mu_c \xi^p \eta^q \nabla_a \nu_d - {\alpha^{cd}}_{pq} \mu_c \nu_d \eta^q \nabla_a \xi^p - {\alpha^{cd}}_{pq} \mu_c \nu_d \xi^p \nabla_a \eta^q \end{aligned}$$

(Exercise 75. Verify that this ∇_a is a derivative operator, and that $\nabla_a g_{bc} = 0$.) Uniqueness follows from the observation that, if ∇'_a is a derivative operator with $\nabla'_a g_{bc} = 0$, then, from (52) and (54), the action of ∇'_a and ∇_a coincide in each of the steps above. The ∇_a of Theorem 12 is called the derivative operator defined by g_{ab}. Whenever we have a manifold with metric, and a derivative operator ∇_a appears without a statement to the contrary, it is to be assumed that ∇_a is the derivative operator of Theorem 12. As a consequence of Theorem 12, indices of tensor fields can be raised and lowered either before or after the application of ∇_a, with the same result. That is, we have

$$\nabla_a {\alpha_m}^{cd} = g^{dp} \nabla_a {{\alpha_m}^c}_p$$

Given any tensor field, e.g., ${\alpha_b}^{cd}$, we write $\nabla^a {\alpha_b}^{cd}$ for $g^{am}\nabla_m {\alpha_b}^{cd}$. That is, raising and lowering of indices extends consistently to the index on the derivative operator.

Example 34. Let g_{ab} be a metric on M, and Ω a scalar field on M which is everywhere positive. Set $g'_{ab} = \Omega^2 g_{ab}$. (Exercise 76. Prove that g'_{ab} is a metric on M. Let ∇_a and ∇'_a be the derivative operators, respectively, with respect to g_{ab} and g'_{ab}. Then ∇_a and ∇'_a are related by (48). We find an expression for $\gamma^m{}_{ab}$. We have

$$0 = \nabla'_a g'_{mc} = \nabla_a g'_{bc} + \gamma^m{}_{ab} g'_{mc} + \gamma^m{}_{ac} g'_{bm}$$
$$= \nabla_a(\Omega^2 g_{bc}) + \Omega^2 \gamma^m{}_{ab} g_{mc} + \Omega^2 \gamma^m{}_{ac} g_{bm}$$
$$= 2\Omega g_{bc} \nabla_a \Omega + \Omega^2 \gamma^m{}_{ab} g_{mc} + \Omega^2 \gamma^m{}_{ac} g_{bm}$$

Solving for $\gamma^m{}_{ab}$,

$$\gamma^m{}_{ab} = -\frac{1}{2}\Omega^{-1} g^{mn}(g_{na}\nabla_a\Omega + g_{nb}\nabla_a\Omega - g_{ab}\nabla_n\Omega \tag{69}$$

The change of metric, $g'_{ab} = \Omega^2 g_{ab}$ is called a *conformal transformation.*)
Exercise 77. Combining Examples 31 and 34, verify that, under a conformal transformation, the Riemann tensors of ∇'_a and ∇_a are related by

$$R'_{abc}{}^d = R_{abc}{}^d + \Omega^{-1} g_{c[b}\nabla_{a]}\nabla^d\Omega - \Omega^{-1}\delta^d{}_{[b}\nabla_{a]}\nabla_c\Omega + 2\Omega^{-2}(\nabla_c\Omega)\delta^d{}_{[b}\nabla_{a]}\Omega$$
$$- 2\Omega^{-2}(\nabla^d\Omega)g_{c[b}\nabla_{a]}\Omega - \Omega^{-2}g_{c[a}\delta_{b]}{}^d(\nabla^m\Omega)(\nabla_m\Omega) \quad \text{(raise, lower with gab)}$$

Exercise 78. Find an example of two distinct metrics on a manifold which determine the same derivative operator. Find metrics of different signature which define the same derivative operator.
Exercise 79. Where did we use invertibility of g_{ab} in Theorem 12? Find an example of a symmetric h_{ab} (not invertible) annihilated by two distinct derivative operators.

14. Curvature Defined by a Metric

We have seen in Sect. 13 that a metric g_{ab} on a manifold leads to a unique derivative operator ∇_a, and in Sect. 12 that a derivative operator on a manifold leads to a Riemann tensor $R_{abc}{}^d$. Hence, a Riemann tensor is associated with any metric. In this case, the Riemann tensor has some further properties which do not hold for Riemann tensor of an arbitrary derivative operator. Furthermore, one has, in the presence of a metric, an important tool for analyzing the Riemann tensor, namely, the ability to raise and lower indices. In this section, we consider the Riemann tensor in Riemannian geometry.

Let M be a manifold, g_{ab} a metric on M, ∇_a the corresponding derivative operator, and $R_{abc}{}^d$ the corresponding Riemann tensor. Then

$$0 = \nabla_{[a}\nabla_{b]}g_{cd} = \frac{1}{2}R_{abc}{}^m g_{md} + \frac{1}{2}R_{abd}{}^m g_{cm} \tag{70}$$

where the first equality follows from $\nabla_a g_{bc} = 0$, and the second from Eqn. (61). Eqn. (70) merely states that $R_{abcd} = g_{md}R_{abc}{}^m$ is antisymmetrical in "c, d". Combining this conclusion with (62) and (63), we have the following algebraic conditions on the Riemann tensor:

$$R_{abcd} = R_{[ab][cd]} \tag{71}$$

$$R_{[abc]d} = 0 \tag{72}$$

These, in fact, exhaust the algebraic conditions on the Riemann tensor in this case. It is convenient, however, to express them in a somewhat different form. Writing (72) explicitly, using (71), we have

$$R_{abcd} + R_{bcad} + R_{cabd} = 0$$

(Exercise 80. Why are there three terms on the left instead of six as in the last equation at the top of page 31?) Using (71) again,

$$R_{abcd} - R_{bcda} - R_{acbd} = 0$$

Now antisymmetrize this equation over "b, c, d,". The second term on the left vanishes by (72), while the first and third terms are equal. Hence, we have

$$R_{a[bcd]} = 0 \tag{73}$$

The equation we are after is derived from (71), (72), and (73) using the figure below.

At each vertex of the octahedron, there appears a Riemann tensor with its indices in a certain order. Given any face of the octahedron, the sum of the three tensors on the vertices of that triangle vanishes, by (71), (72). For example, the sum corresponding to the rear face in the top half of the octahedron is $R_{bdca} + R_{cdab} + R_{dacb}$, which, by (71) is equal to $R_{dbac} + R_{dcba} + R_{dacb}$, which, by (73), vanishes. Now add the four expressions obtained from the four faces in the bottom half of the figure, and

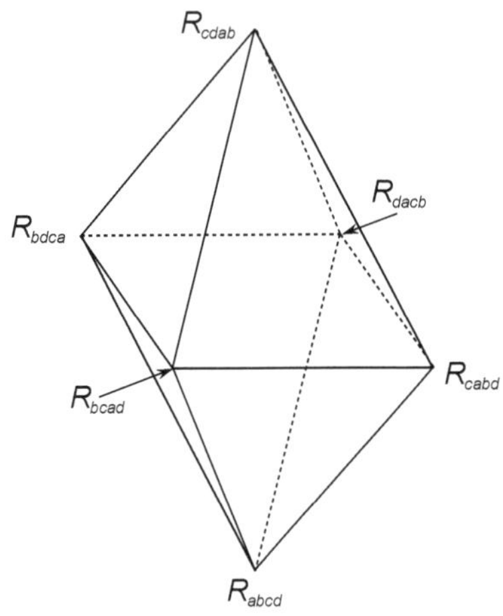

subtract from this the sum of the expressions obtained from the four faces in the top half of the figure. Since each expression is zero, the result is zero. But, when this quantity is written out, all terms which arise from the square "equator" in the figure will cancel out. (Each such term will appear four times, twice (with plus signs) because of the sum over the top faces.) Hence, this quantity, which is zero, becomes $4R_{abcd} - 4R_{cdab}$. We conclude that

$$R_{abcd} = R_{cdab} \tag{74}$$

Eqns. (71) and (74) can now be combined into a single equation:

$$R_{abcd} = R_{[cd][ab]} \tag{75}$$

Thus, as a complete set of algebraic conditions on the Riemann tensor, we may take (72) and (75).

Example 35. Let R_{abcd} be a tensor field satisfying (75). We show that this R_{abcd} satisfies (72) if and only if it satisfies

$$R_{[abcd]} = 0 \tag{76}$$

Clearly, (72) implies (76). Next, suppose (76). Then the expression

$$R_{abcd} - R_{abdc} + R_{adbc} - R_{dabc}$$

when antisymmetrized over "a, b, c", vanishes. Using (75), this is the same as $R_{abcd} + R_{abcd} - R_{dabc} - R_{dabc}$. Hence,

$$R_{[abc]d} - R_{d[abc]} = 0$$

But, from (75), these two terms are equal. Hence, we have (72).
Example 36. We show that the complete set of algebraic conditions on the Riemann tensor can be expressed by the single condition

$$R_{abcd} = R_{[cd][ab]} + R_{[abcd]} \tag{77}$$

That is, we show that a tensor field R_{abcd} satisfies (72) and (75) if and only if it satisfies (77). Clearly, (72) and (75) imply (77). To prove the converse, antisymmetrize (77) over "a, b, c, d". All three terms become $R_{[abcd]}$, whence this quantity must vanish. Discarding this term in (77), we obtain (75). But, by Example 35, (75) and (76) imply (72).
Exercise 81. One might think he could obtain an additional algebraic condition on R_{abcd} by noting that, from Bianchi's identity,

$$\nabla_{[a}\nabla_{b}R_{cd]ef} = 0$$

and that, from (61), the left side can be expressed in term of the Riemann tensor with no derivatives. Check that this yields nothing new.

In Riemannian geometry, a vector field ξ^a is said to be a *Killing vector* if $\mathcal{L}_\xi g_{ab} = 0$. By (52), ξ^a is a Killing vector if and only if

$$\nabla_{(a}\xi_{b)} = 0 \tag{78}$$

The Killing vectors clearly form a vector space. By (31), they form a Lie algebra. (It is a theorem that, if M is connected, the vector space of Killing vectors on M, g_{ab} is finite -dimensional.)
Exercise 82. The remarks surrounding Eqn. (66) suggest that, if ξ^a is a Killing vector for M, g_{ab}, then ξ^a is an affine collineation for ∇_a. Prove it.

Additional curvature tensor fields can be obtained by taking contractions of the Riemann tensor. The tensor field

$$R_{ab} = R_{amb}{}^{m} \tag{79}$$

is called the *Ricci tensor* (from (75), $R_{ab} = R_{(ab)}$), and the scalar field

$$R = R_{m}{}^{m} \tag{80}$$

the *scalar curvature*. Note that the result of contracting any two indices of the Riemann tensor is either zero or $\pm R_{ab}$, and that the result of contracting all indices of the Riemann tensor is either zero or $\pm R$. Thus, the Ricci tensor and scalar curvature are the only tensor fields obtainable by contracting a single Riemann tensor. The Bianchi identity, (64), implies a differential identity on R_{ab} and R. Contracting $\nabla_{[a}R_{bc]}{}^{de}$ first over "b" and "d", and then over "c" and "e", we obtain

$$\nabla_m(R^{am} - \frac{1}{2}Rg^{am}) = 0 \tag{81}$$

One should think of the Ricci tensor and scalar curvature as "parts" of the entire Riemann tensor. This remark is made more explicit by defining the *Weyl tensor* by the equation (n = dimension $M > 2$)

$$R_{abcd} = C_{abcd} + \frac{2}{n-2}(g_{a[c}R_{d]b} - g_{b[c}R_{d]a}) - \frac{2}{(n-1)(n-2)}Rg_{a[c}g_{d]b} \tag{82}$$

Clearly, C_{abcd} has all the algebraic symmetries of the Riemann tensor. Furthermore, contracting (82), we see that all contractions of C_{abcd}, e.g., $C_{mab}{}^{m}$ vanish.

Exercise. 83 Let g_{ab} be a metric, Ω a positive scalar field, and $g'_{ab} = \Omega^2 g_{ab}$ another metric. (That is, we have a conformal transformation.) Let $C_{abc}{}^{d}$ and $C'_{abc}{}^{d}$ be the corresponding Weyl tensors. Prove that $C_{abc}{}^{d} = C'_{abc}{}^{d}$. (This fact is normally expressed by saying that the Weyl tensor is conformally invariant.)

Exercise 84. Let ξ^a be a Killing vector. Prove that $\mathcal{L}_\xi R_{abcd} = 0$. Prove, similarly, that the Lie derivative of the Ricci tensor, scalar curvature, and Weyl tensor in the ξ–direction are all zero.

Exercise 85. Let R_{abcd} be a tensor field satisfying (75) and (72), and set $P_{abcd} = R_{c(ab)d}$ Show that P_{abcd} satisfies

$$P_{abcd} = P_{(cd)(ab)} \quad P_{(abc)d} = 0$$

Find an expression for R_{abcd} in terms of P_{abcd}. Finally, show that, given any tensor field P_{abcd} satisfying the condition above, your expression yields an R_{abcd} satisfying (75) and (72).

15. Smooth Mappings: Action on Tensor Fields

Let M and M' be manifolds. Recall that a mapping $\psi : M \to M'$ is said to be *smooth* if, for every smooth function f' on M', the function $f = f' \cdot \psi$ on M is also smooth.

Example 37. Let $M = R^n$, $M' = R^{n'}$, and let k be a nonnegative integer less than or equal to n and less than or equal to n'. Set $\psi(x_1, \ldots, x^n) = (x^1, \ldots, x^k, 0, \ldots, 0)$. Then, if $k = 0$, all of M is mapped to the origin of M'. The smooth mapping ψ is one-to-one if and only if $k = n$, and onto if and only if $k = n'$. Finally, ψ is a diffeomorphism if and only if $k = n = n'$.

A natural question to ask is: Under what conditions does a smooth mapping from one manifold to another carry a tensor at a point, or a tensor field, form one manifold to the other? This question has a simple and complete answer, which we now obtain.

We begin with tensors at a point. Let $\psi : M \to M'$, a point of M' be smooth, let p be a point of M, and let $p' = \psi(p)$, a point of M'. Consider any contravariant vector ξ^a at the point p. Then, for any smooth function f' on M', set $\eta'(f') = \xi(f)$, where we have set $f = f' \cdot \psi$, and where $\xi(f)$ is the directional derivative as described in Sect. 3. Thus, η' associates, with any smooth function f' on M', a real number, $\eta'(f')$. We verify that this $\eta'(f')$ satisfies DD1–DD3 of Sect. 3. Properties DD1 and DD3 are obvious. To check DD2, note that, if $f' = g'h'$, then $f = gh$. Furthermore, $f'(p') = f(p)$, $g'(p') = g(p)$, and $h'(p') = h(p)$. The verification of DD2 is now easy: $\eta'(g'h') = \xi(gh) = g(p)\xi(h) + h(p)\xi(g) = g'(p')\eta'(h') + h'(p')\eta(g')$. Thus, a contravariant vector ξ at the point p of M defines a contravariant vector η' at the point p' of M'. We write $\eta'^a = \vec{\psi}_p \xi^a$. Thus, $\vec{\psi}_p$ is a (clearly) linear mapping from the vector space of contravariant vectors at p of M to the vector space of contravariant vectors at p' of M'.

Next, let μ'_a be a covariant vector in M' at p'. Then, if ξ^a is any contravariant vector in M at p, consider the number $\mu'_a(\vec{\psi}_p \xi^a)$. (This is well-defined, for $\vec{\xi}_p \xi^a$ is a contravariant vector in M' at p', while μ'_a is a contravariant vector in M' at p'.) The number $\mu'_a(\vec{\psi}_p \xi^a)$ is clearly linear in ξ^a.

Hence, we have just described a linear mapping from contravariant vectors in M at p to real numbers. In other words, we have just described a covariant vector in M at p. We write this covariant vector as $\underset{\leftarrow}{\psi_p}\,\mu'_a$, so $\underset{\leftarrow}{\psi_p}\,\mu'_a$, is defined by the equation $\xi^a(\underset{\leftarrow}{\psi_p}\,\mu'_a) = \mu'_a(\vec{\psi}_p\xi^a)$ for all ξ^a in M at p. Thus, $\underset{\leftarrow}{\psi_p}$ is a (clearly) linear mapping from the vector space of covariant vectors in M' at p' to the vector space of covariant vectors in M at p.

Finally, we extend to tensors – rather than just vectors – at a point. Let $\alpha^{a\ldots c}$ be a contravariant tensor (= tensor all of whose indices are contravariant) in M at p. Then, for any vectors $\mu'_a, \ldots, \nu'_c$ in M' at p', the right side of

$$(\vec{\psi}_p\alpha^{a\ldots c})\mu'_a \ldots \nu'_c = \alpha^{a\ldots c}(\underset{\leftarrow}{\psi_p}\,\mu'_a)\ldots(\underset{\leftarrow}{\psi_p}\,\nu'_c) \tag{83}$$

is multilinear in $\mu'_a, \ldots, \nu'_c$. Hence, Eqn. (83) defines a contravariant tensor, $\vec{\psi}_p\alpha^{a\ldots c}$, at the point p' of M'. Similarly, if $\beta'_{a\ldots c}$ is a covariant tensor in M' at p', and $\xi^a, \ldots, \eta^c$ are vectors in M at p, then the right side of

$$(\underset{\leftarrow}{\psi_p}\,\beta'_{a\ldots c})\xi^a \ldots \eta^c = \beta'_{a\ldots c}(\vec{\psi}_p\xi^a)\ldots(\vec{\psi}_p\eta^c) \tag{84}$$

is multilinear in $\xi^a, \ldots, \eta^c$, defining a covariant tensor, $\underset{\leftarrow}{\psi_p}\,\beta'_{a\ldots c}$, in M at p. Clearly, the mappings $\vec{\psi}_p$ and $\underset{\leftarrow}{\psi_p}$, as extended above to tensor of arbitrary rank, are linear.

We now have a mapping $\vec{\psi}_p$ from contravariant tensors in M at p to contravariant tensors in M' at p', and a mapping $\underset{\leftarrow}{\psi_p}$ from covariant tensors in M' at p' to covariant tensors in M at p. How do these mappings interact with the tensor operations? We first consider outer product. Let $\alpha'_{a\ldots c}$ and $\beta'_{b\ldots d}$ be at p', and $\xi^a, \ldots, \tau^c, \eta^b, \ldots, \lambda^d$ be at p. Then,

$$\begin{aligned}
&\left[\underset{\leftarrow}{\psi_p}\,(\alpha'_{a\ldots c}\beta'_{b\ldots d})\right]\xi^a \cdots \lambda^d = \alpha'_{a\ldots c}\beta'_{b\ldots d}(\vec{\psi}_p\xi^a)\ldots(\vec{\psi}_p\lambda^d)\\
&= \left[\alpha'_{a\ldots c}(\vec{\psi}_p\xi^a)\ldots(\vec{\psi}_p\tau^c)\right]\left[\beta'_{b\ldots d}(\vec{\psi}_p\eta^b)\ldots(\vec{\psi}_p\lambda^d)\right]\\
&= (\underset{\leftarrow}{\psi_p}\,\alpha'_{a\ldots c})\xi^a \ldots \tau^c\,(\underset{\leftarrow}{\psi_p}\,\beta'_{b\ldots d})\eta^b \ldots \lambda^d
\end{aligned}$$

Evidently, $\underset{\leftarrow}{\psi_p}\,(\alpha'_{a\ldots c}\beta'_{b\ldots d}) = \underset{\leftarrow}{\psi_p}\,(\alpha'_{a\ldots c})\,\underset{\leftarrow}{\psi_p}\,(\beta'_{b\ldots d})$. Similarly for $\vec{\psi}_p$.

The second operation to be considered is contraction. Let $\alpha^{a\ldots bc\ldots d}$ be at p, and $\mu'_a \ldots, \kappa'_b, \nu'_c, \ldots, \rho'_d$ be at p'. Then

$$(\vec{\psi}_p\alpha^{a\ldots d})\mu'_a \ldots \kappa'_b\nu'_c \ldots \rho'_d = \alpha^{a\ldots d}(\underset{\leftarrow}{\psi_p}\,\mu'_a)\ldots(\underset{\leftarrow}{\psi_p}\,\kappa'_b)(\underset{\leftarrow}{\psi_p}\,\nu'_c)\ldots(\underset{\leftarrow}{\psi_p}\,\rho'_d)$$

$$= \vec{\psi}_p\left[\alpha^{a\ldots d}(\underset{\leftarrow}{\psi_p}\,\mu'_a)\ldots(\underset{\leftarrow}{\psi_p}\,\kappa'_b)\right]\nu'_c \ldots \rho'_d$$

where the first equally is just (83), and in the second equality, we are again using (83), but taking $\mu'_a, \ldots, \kappa'_b$ as fixed. Since $\nu'_c, \ldots, \rho'_d$ are arbitrary, we have

$$(\vec{\psi}_p\alpha^{a\ldots d})\mu'_a \ldots \kappa'_b = \vec{\psi}_p\left[\alpha^{a\ldots d}\underleftarrow{\psi_p}\,\mu'_a)\ldots(\underleftarrow{\psi_p}\,\kappa'_b)\right]$$

Finally, noting that every covariant tensor at p' can be written as a sum outer products of vectors, we have

$$(\vec{\psi}_p\alpha^{a\ldots bc\ldots d})\beta'_{a\ldots b} = \vec{\psi}_p\left[\alpha^{a\ldots bc\ldots d}\underleftarrow{\psi_p}\,\beta'_{a\ldots b}\right] \tag{85}$$

Similarly with contravariant and covariant, p and p', etc. reversed.

We summarize:

Theorem 13. Let $\psi : M \to M'$ be a smooth mapping of manifolds. Let p be a point of M, and let $p' = \psi(p)$, a point of M'. Then there are natural, rank-preserving mappings $\vec{\psi}_p$ from contravariant tensors in M at p to contravariant tensors in M' at p', and $\underleftarrow{\psi_p}$ from covariant tensor in M' at p' to covariant tensors in M at p, such that:

1. $$\vec{\psi}_p(\alpha^{a\ldots c} + c\beta^{a\ldots c}) = \vec{\psi}_p\alpha^{a\ldots c} + c\vec{\psi}_p\beta^{a\ldots c}$$
$$\underleftarrow{\psi_p}\,(\gamma'_{a\ldots c} + c\mu'_{a\ldots c}) = \underleftarrow{\psi_p}\,\gamma'_{a\ldots c} + c\underleftarrow{\psi_p}\,(\mu'_{a\ldots c}$$

where c is a constant.

2. $$\vec{\psi}_p(\alpha^{a\ldots c}\beta^{b\ldots d}) = (\vec{\psi}_p(\alpha^{a\ldots c})(\vec{\psi}_p(\beta^{b\ldots s})$$
$$\underleftarrow{\psi_p}\,(\mu'_{a\ldots c}\,\nu'_{b\ldots d}) = \underleftarrow{\psi_p}\,(\mu'_{a\ldots c})(\underleftarrow{\psi_p}\,(\nu'_{b\ldots d}$$

3. $$(\vec{\psi}_p\alpha^{a\ldots bc\ldots d})\mu'_{a\ldots b} = \vec{\psi}_p\left[\alpha^{a\ldots bc\ldots d}\,\underleftarrow{\psi_p}\,\mu'_{a\ldots b}\right]$$
$$(\underleftarrow{\psi_p}\,(\mu'_{a\ldots bc\ldots d})\beta^{a\ldots b} = \underleftarrow{\psi_p}\left[\nu'_{a\ldots bc\ldots d}\vec{\psi}_p\beta^{a\ldots b}\right]$$

Theorem 13 is essentially the whole story regarding tensors at a point. There are no other "natural" mappings available, i.e., no reasonable way in general to carry a covariant tensor in M at p to a covariant tensor in M' at p'. However, when we pass to tensor fields, some more things are true.

We begin with contravariant fields on M. Let $\xi^{a\ldots c}$ be such a tensor field. Then, for each point p of M, $\vec{\psi}_p\,\xi^{a\ldots c}(p)$ is a tensor at the point $p' = \psi(p)$ of M! Does this define a tensor field on M'? The answer is no. For one thing, there may be distinct points p and q of M such that $\psi(p) = \psi(q) = p'$ (i.e., ψ may not be one-to-one; Example 37) and such that $\vec{\psi}_p\xi^{a\ldots c}(p)$ and $\vec{\psi}_q\xi^{a\ldots c}(q)$

(both tensors in M' at p') are not equal. Which tensor at p' should we choose to get a tensor field on M'? Furthermore, there may be points of M' which are not the image under of any point of M (i.e., ψ may not be onto; Example 37). A contravariant tensor field on M defines no tensor whatever at such points of M'. Clearly, this is no good.

We try it the other way. Let $\mu'_{a...c}$ be a tensor field on M'. Then, for each point p of M $\underset{\leftarrow}{\psi}_p$ $(\mu'_{a...c}(p')$ is a covariant tensor at p. Since this is true for each point p of M, we have defined a covariant tensor field on M. Thus, with each covariant tensor field $\mu'_{a...c}$ on M' we associate a covariant tensor field, $\underset{\leftarrow}{\psi}\ \mu'_{a...c}$ on M. (We have not yet proven that $\underset{\leftarrow}{\psi}\ \mu'_{a...c}$ is smooth if $\mu'_{a...c}$ is.) Evidently, the linearity and outer-product behavior from Theorem 13 extend from tensors at a point to tensor fields.

The mapping $\underset{\leftarrow}{\psi}$ is not very interesting unless it takes smooth tensor fields to smooth tensor fields. We now show that it does. First, scalar fields (covariant tensor fields of rank zero). If f' is a smooth scalar field on M', then $\underset{\leftarrow}{\psi} f' = f = f' \cdot \psi$ is also smooth, by the definition of a smooth mapping. We next show that

$$D(\underset{\leftarrow}{\psi} f')_a = \underset{\leftarrow}{\psi}[D(f')_a] \tag{86}$$

Let ξ^a be a vector in M at p. Then

$$\begin{aligned}\xi^a D(\underset{\leftarrow}{\psi} f')_a\big|_p &= \xi^a D(f)_a\big|_p = \xi(f) = [\vec{\psi}_p \xi](f') \\ &= (\vec{\psi}_p \xi^a) D(f')_a\big|_p = \xi^a \underset{\leftarrow}{\psi}\ [D(f')_a]\big|_p\end{aligned}$$

where the second equality is the definition of the gradient, the third is the definition of $\vec{\psi}_p$, the fourth is the definition of the gradient, and the fifth is the definition of $\underset{\leftarrow}{\psi}$. That is to say, Eqn. (86), evaluated at any point p of M and contracted with any contravariant vector in M at p, holds. Hence, Eqn. (86) holds. Now let μ'_a be a covariant vector field in M'. Then, for each point p' of M', there exists an open subset O' of M', containing p', and smooth functions $f', g' \ldots, h', k'$ on M' such that, in O',

$$\mu'_a = f'\ D(g')_a + \ldots + h'\ D(k')_a \tag{87}$$

(Proof: Introduce a chart, use components, and repeat the discussion of page 20.) Hence, from (86)

$$\underset{\leftarrow}{\psi}\ \mu'_a = (\underset{\leftarrow}{\psi}\ f')\ D\ (\underset{\leftarrow}{\psi}\ g')_a + \ldots + (\underset{\leftarrow}{\psi}\ h')D\ (\underset{\leftarrow}{\psi} k')_a \tag{88}$$

Since the right side of (88) is smooth, we conclude that $\underset{\leftarrow}{\psi}\ \mu'_a$ is smooth in $\vec{\psi}^{-1}[O']$. But every point of M' is contained in such an O', and so $\underset{\leftarrow}{\psi}\ \mu'_a$ is smooth on M. Finally, let $\alpha'_{a...c}$ be a tensor field on M' of arbitrary rank. Then every point of M' is contained in an open set in which

$$\alpha'_{a...c} = \mu'_a \cdots k'_c + \cdots + \nu'_a \cdots \rho'_c \tag{89}$$

Then, in ψ^{-1} of this open set,

$$\underset{\leftarrow}{\psi}\, \alpha'_{a...c} = (\underset{\leftarrow}{\psi}\, \mu'_a) \cdots (\underset{\leftarrow}{\psi}\, \kappa'_c) + \cdots + (\underset{\leftarrow}{\psi}\, \nu'_a) \cdots (\underset{\leftarrow}{\psi}\, \rho'_c)$$

Since the right side is smooth, so is the left, Since every point of M' is contained in such an open set, $\underset{\leftarrow}{\psi}\, \alpha'_{a...c}$ is a smooth tensor field on M.

Thus, $\underset{\leftarrow}{\psi}$ is a rank-preserving mapping from smooth covariant tensor fields on M' to smooth covariant tensor fields on M. Finally, we drove that "$\underset{\leftarrow}{\psi}$ commutes with exterior differentiation:"

$$\underset{\leftarrow}{\psi}\, D(\omega')_{ma...c} = D(\underset{\leftarrow}{\psi}\, \omega')_{ma...c} \tag{90}$$

For ω' a 0−form, (90) is just (86). For ω' a 1−form, we take the exterior derivative of (87),

$$D(\omega')_{ma} = D(f')_{[m} D(g')_{a]} + \cdots + D(h')_{[m} D(k')_{a]}$$

and apply $\underset{\leftarrow}{\psi}$,

$$\begin{aligned}\underset{\leftarrow}{\psi}\, D(\omega')_{ma} &= D(\underset{\leftarrow}{\psi}\, f')_{[m}\, D(\underset{\leftarrow}{\psi}\, g')_{a]} + \cdots + D(\underset{\leftarrow}{\psi}\, h')_{[m}\, D(\underset{\leftarrow}{\psi}\, k')_{a]} \\ &= D(\underset{\leftarrow}{\psi}\, f'\, D(\underset{\leftarrow}{\psi}\, g') + \cdots + \underset{\leftarrow}{\psi}\, (h')\, D(\underset{\leftarrow}{\psi}\, k'))_{ma} \\ &= D(\underset{\leftarrow}{\psi}\, \omega')_{ma}\end{aligned}$$

For $\omega_{a...c}$ a p−form, we take the exterior derivative of (89),

$$\begin{aligned}D(\alpha')_{ma...c} &= D(\mu')_{ma} \cdots \kappa'_c + \cdots + \mu'_a \cdots D(\kappa')_{mc} \\ &+ \cdots + D(\nu')_{ma} \cdots \rho'_c + \cdots + \nu'_a \cdots D(\rho')_{mc}\end{aligned}$$

and apply $\underset{\leftarrow}{\psi}$,

$$\begin{aligned}\underset{\leftarrow}{\psi}\, D(\alpha')_{ma...c} &= D(\underset{\leftarrow}{\psi}\, \mu')_{ma} \cdots \underset{\leftarrow}{\psi}\, \kappa'_c + \cdots + \underset{\leftarrow}{\psi}\, \mu'_a \ldots D(\underset{\leftarrow}{\psi}\, \kappa')_{mc} \\ &+ \cdots + D(\underset{\leftarrow}{\psi}\, \nu')_{ma} \cdots \underset{\leftarrow}{\psi}\, \rho'_c + \cdots + \underset{\leftarrow}{\psi}\, \nu'_a \cdots D(\underset{\leftarrow}{\psi}\, \rho')_{mc} \\ &= D(\underset{\leftarrow}{\psi}\, \alpha')_{ma\cdots c}\end{aligned}$$

To summarize,

Theorem 14. The mapping $\underset{\leftarrow}{\psi}_p$ of Theorem 13 extends to a mapping, $\underset{\leftarrow}{\psi}$ from (smooth) covariant tensor fields on M' to (smooth) covariant tensor fields on M, with $\underset{\leftarrow}{\psi}$ satisfying the following:

1. $\underset{\leftarrow}{\psi}(\gamma'_{a\cdots c} + \mu'_{a\cdots c}) = \underset{\leftarrow}{\psi}(\gamma'_{a\cdots c}) + \underset{\leftarrow}{\psi}(\mu'_{a\cdots c})$.
2. $\underset{\leftarrow}{\psi}(mu'_{a\cdots c}\nu'_{b\cdots d}) = \underset{\leftarrow}{\psi}(\mu'_{a\cdots c})\underset{\leftarrow}{\psi}(\nu'_{b\cdots d})$.
3. For$\omega_{a\cdots}$ a $p-$form, $\underset{\leftarrow}{\psi}\ D(\omega)_{ma\cdots} = D(\underset{\leftarrow}{\psi}\ \omega)_{ma\cdots c}$

We summarize what exists in the following table:

	Tensor at a point			Tensor fields		
$\psi : M \to M'$	contra-variant	co-variant	mixed	contra-variant	co-variant	mixed
From M to M'	$\vec{\psi}_p$	—	—	—	—	—
From M' to M	—	$\underset{\leftarrow}{\psi}_p$	—	—	$\underset{\leftarrow}{\psi}$	—

Finally, suppose that $\psi : M \to M'$ is a diffeomorphism, so $\psi^{-1} : M' \to M$ also exists and is smooth. Let $\alpha'^{a\cdots c}{}_{b\cdots d}$ be a tensor field on M'. Then, for any vector fields $\mu_a \cdots \nu_c$ on M, the right side of

$$\beta^{a\ldots c}{}_{b\cdots d}\,\mu_a \cdots \nu_c = \underset{\leftarrow}{\psi}\left[\alpha'^{a\cdots c}{}_{b\cdots d}\,(\underset{\leftarrow}{\psi^{-1}}\mu_a)\cdots(\underset{\leftarrow}{\psi^{-1}}\nu_c)\right]$$

is linear in $\mu_a \cdots \nu_c$. Hence, we define a tensor field, $\beta^{a\ldots c}{}_{b\cdots d}$ on M. Clearly, a diffeomorphism between two manifolds defines a one-to-one correspondence between the tensor fields on one manifold and those on the other, where this correspondence preserves everything (rank, symmetry, number of contravariant and covariant indices, commutes with addition, outer product, index substitution, and contraction, commutes with Lie and exterior differentiation, etc.). This is to be expected. A diffeomorphism between manifolds means they are "identical as manifolds", while tensor fields exist on things with a manifold structure.

Exercise 86. Determine $\vec{\psi}_p$ and $\underset{\leftarrow}{\psi}_p$ for the mappings of Example 37.

Exercise 87. Why don't we write $\underset{\leftarrow}{\psi}_{p'}$ instead of $\underset{\leftarrow}{\psi}_p$?

Exercise 88. Why don't we ask what happens to Lie derivatives under $\underset{\leftarrow}{\psi}$?

Exercise 89. Find an example of a smooth onto mapping ψ such that $\vec{\psi}_p$ is an isomorphism (of vector spaces) for every p, but such that ψ is not a diffeomorphism.

Exercise 90. Let $\psi : M \to M'$ be smooth, so $\varphi \cdot \psi : M' \to M''$ be smooth, so $\varphi \cdot \psi : M \to M''$ is smooth. Prove that $|\underleftarrow{\psi}| \cdot |\underleftarrow{\varphi}| = \underleftarrow{(\varphi \cdot \psi)}$. Let p be a point of M, $p' = \psi(p)$, and $p'' = \varphi(p')$. Prove that $\underleftarrow{(\varphi \cdot \psi)}_p = \underleftarrow{\psi}_p \cdot \underleftarrow{\varphi}_{p'}$ and $\overrightarrow{(\varphi \cdot \psi)}_p = \vec{\phi}_{p'} \cdot \vec{\psi}_p$.

Exercise 90a. Find an example in which g'_{ab} is a metric on M', but $\underleftarrow{\psi}\, g'_{ab}$ is not a metric on M.

Exercise 91. Let $\psi : M \to M'$ be smooth. Show that there is no natural way, in general, to take a derivative operator from M to M', or to take a derivative from M' to M.

Exercise 92. Let $\psi : M \to M'$ be smooth. Let ξ^a be a vector field on M such that, for every point p of M, $\vec{\psi}_p\, \xi^a(p) = 0$. Prove that, for every $\alpha'_{a\cdots c}$ on M', $\mathcal{L}_\xi\, (\underleftarrow{\psi}\, \alpha'_{a\cdots c}) = 0$

Exercise 93. Prove that $\vec{\psi}_p$ (resp. $\underleftarrow{\psi}_p$) is onto if and only if $\underleftarrow{\psi}_p$ (resp. $\vec{\psi}_p$) is one-to-one.

Exercise 94. Let g_{ab} be a metric on M, and g'_{ab} a metric on M'. A diffeomorphism $\psi : M \to M'$ is said to be an *isometry* if $\underleftarrow{\psi}\, g'_{ab} = g_{ab}$. Prove that, if ψ is an isometry, then $\underleftarrow{\psi}\, R'_{abcd} = R_{abcd}$.

Exercise 95. Express the action of $\vec{\psi}_p$ in terms of components with respect to charts. (The mapping $\vec{\psi}_p$ is often called the derivative of ψ at p.)

16. Bundles

We have seen several situations in which, at each point of a manifold, there sits a "space". For example, at each point p of a manifold, we have the space of contravariant vectors at p. There exists a class of mathematical objects, called fibre bundles, in which structure of this general type is isolated. A certain class of fibre bundles, the smooth vector bundles, are rather simpler to define, and, furthermore, suffice for essentially all applications in differential geometry. In this section, we define a smooth vector bundle, and derive a few of its elementary properties.

A smooth vector bundle consists, firstly, of a manifold B (called the *bundle space*), a manifold M (called the *base space*), and a smooth mapping $\pi : B \to M$ (called the *projection*). For each point p of M, the subset $\pi^{-1}[p]$ (i.e, the set of all points P of B such that $\pi(P) = p$) of B is called the *fibre* over p. It is convenient to represent this situation by a figure with the bundle space drawn over the base space, and the projection mapping taking each point of B "vertically downward" to the point of M directly below it:

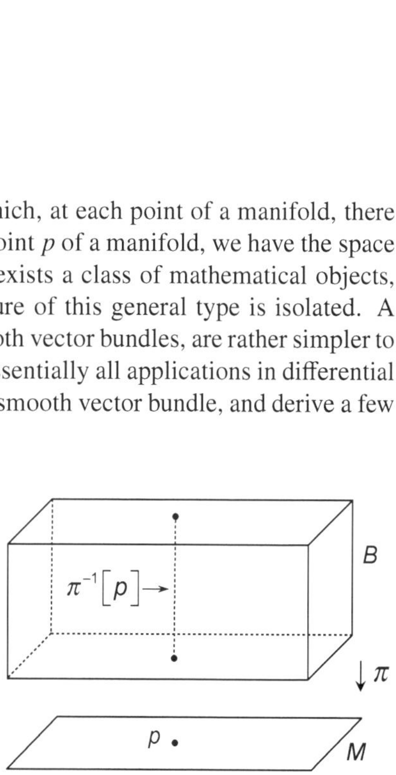

In terms of the figure, the fibre over p is the vertical line in B directly above p. We next require that each fibre have the structure of a k–dimensional vector space, where the non-negative integer k is fixed (i.e., does not vary from one fibre to another). In more detail, this means that linear combinations (with real coefficients) of elements of a single fibre are defined, that these combinations are again elements of that fibre, and that all the axioms for a k–dimensional vector space are satisfied by this operation on this one fibre. (We cannot "add" elements of different fibres.)

This is not quite the end of the definition. We still wish to ensure that "locally, the bundle space B is a product of the base space and a k–dimensional vector space." This idea is formulated as follows. Let U be a "sufficiently small" open subset of M. Then $\pi^{-1}[U]$ (i.e., the subset of B consisting of all points of B which are mapped into U by π) should have "exactly the same structure" as $U \times V^k$, where V^k is a k–dimensional vector space (and, therefore, a k–dimensional manifold, so $U \times V^k$ is a product of manifolds). That is, we require that there exists a diffeomorphism $\psi : U \times V^k \to \pi^{-1}[U]$ which is "structure preserving". What is there to be preserved? For each point p of U, denote by $\{p\}$ the subset of $U \times V^k$ consisting of all pairs (p, v), for all elements v of V^k. We require, firstly, that ψ map each subset of $U \times V^k$ of the form $\{p\}$ onto the fibre over p. In other words, ψ should be "fibre preserving". That is, we require that the point $\psi(p, v)$ of B lie in the fibre over p. That is, we require that $\pi \cdot \psi(p, v) = p$. There is one more piece of structure to be preserved by ψ. Each subset $\{p\}$ of $U \times V^k$ is, of course, a k–dimensional vector space. But each fibre in B is also a k–dimensional vector space. Since ψ takes $\{p\}$ to the fibre over p, we might as well require that that ψ also preserve the vector-space structures. Explicitly, we require that, for v and w in V^k and c a number, $\psi(p, v + cw) = \psi(p, v) + c\,\psi(p, w)$. (Note that the right side is well-defined, for $\psi(p, v)$ and $\psi(p, w)$ lie in the same fibre, namely, the fibre over p.) Thus, the "local product" condition is that, for every point q of M, there exists an open subset U of M and a diffeomorphism $\psi : U \times V^k \to \pi^{-1}[U]$ which is fibre-preserving and vector-space structure preserving.

We summarize this discussion with the definition. A (*smooth*) *vector bundle* consists of a manifold B, a manifold M, a smooth mapping $\pi : B \to M$, a non-negative integer k, and a k–dimensional vector space structure on $\pi^{-1}[p]$ for each point p of M such that: for each point q of M there is an open subset U of M, containing q, and a diffeomorphism $\psi : U \times V^k \to \pi^{-1}[U]$ satisfying i) $\pi \cdot \psi(p, v) = p$, and ii) $\psi(p, v + cw) = \psi(p, v) + c\psi(p, w)$.

Example 38. Let M be a manifold, and V^k a k–dimensional vector space. Set $B = M \times V^k$, and let $\pi : B \to M$ be defined by $\pi(p, v) = p$. Then the fibre over a point p of M is the collection of all points (p, v) of B for all v in V^k. Clearly, these fibres have the structure of a k–dimensional vector space. It is also clear that this is a fibre bundle. (Not only is B a "local product", but also a "global product".)

Example 39. Let $M = R^1$, so a point of M is a real number x. Let $B = R^2$, so a point of B is a pair, (y, z), of real numbers. Let π be defined by $\pi(y, z) = y^3$, a point of M. Then the fibre over the point x of M is the subset of B consisting of pairs $(x^{1/3}, z)$, for all z. These fibres have an obvious one-dimensional vector space structure, i.e., $(x^{1/3}, z) + c(x^{1/3}, w) = (x^1/3, z + cw)$. However, this is not a vector bundle, for B is not a local product near the point $x = 0$ of M. (Exercise 96. Verify that the above structure does not define a vector bundle.)

We now obtain a few properties of vector bundles. Note, firstly, that the projection π must be onto M, for, if a point q of M were not in the image of B

under π, then $\pi^{-1}[q]$ would be the empty set. But the empty set cannot have the structure of a k–dimensional vector space. Next, note, that, if the base space M is n–dimensional, and if the fibres are k–dimensional vector spaces, then the bundle space B must have dimension $(n+k)$. This is immediate from the local product property and the fact that, for products of manifolds, the dimensions add. (That is, $U \times V^k$ has dimension $(n + k)$.) The next property is slightly more subtle. Let P be a point of B, and let $\pi(P) = p$, a point of M. Then $\vec{\pi}_P$ is a linear mapping from the $(n + k)$–dimensional vector space of contravariant vectors in M at p. We want to describe this mapping $\vec{\pi}_P$ in more detail. Let U and $\psi : U \times V^k \to \pi{-}1[U]$ be as in the local product part of the definition, and let v be the element of V^k such that $\psi(p, v) = P$. Let $\alpha : U \to U \times V^k$ be defined by $\alpha(q) = (q, v)$, and $\beta : V^k \to U \times V^k$ defined by $\beta(w) = (p, w)$. Then $\psi \cdot \beta : V^k \to B$ maps V^k to the fiber over p. Hence, $\pi \cdot \psi \cdot \beta : V^k \to M$ maps all of V^k to the single point of p of M. Hence, any contravariant vector in the manifold V^k at the point v is mapped, by $(\pi \cdot \vec{\psi} \cdot \beta)_v$ to the zero contravariant vector in M at p. On the other hand, $\pi \cdot \psi \cdot \alpha : U \to U$ is just the identity mapping. Hence, $(\pi \cdot \vec{\psi} \cdot \alpha)_p$ maps a contravariant vector in U at p to that same contravariant vector in U at p. Thus, the $(n + k)$–dimensional vector space of contravariant vectors in B at P has a k–dimensional subspace consisting of vectors which get mapped to zero, and a n–dimensional subspace consisting of vectors which (except for the zero one) do not get mapped to zero. That is, $\vec{\pi}_P$, a linear mapping from a $(n + k)$–dimensional space to an n–dimensional space, is onto, and annihilates a certain k–dimensional subspace of the vector space of contravariant vectors in B at P. Since vector bundles are "locally products", things like $\vec{\pi}_P$, which are local, have exactly the same properties that they have for products.

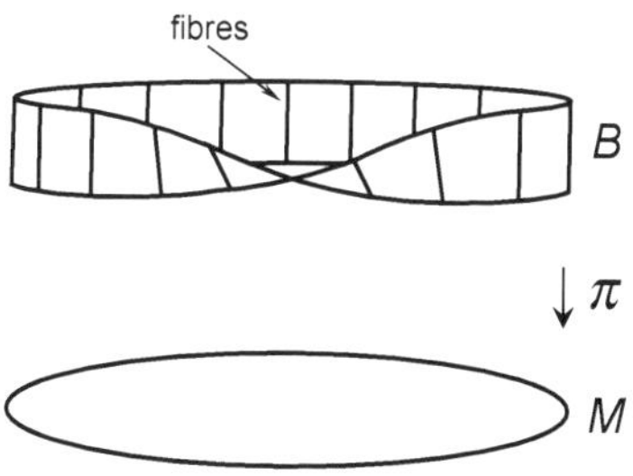

Example 40. If, in Example 38, we let $M = S^1$ (the circle), and $k = 1$ (so V^k is a one-dimensional vector space, i.e., a line), then the bundle space B is, a manifold, the cylinder, $S^1 \times R^1$. But there is another bundle with base space S^1, and with $k = 1$, which is not a simple product. Let B consist of pairs, (θ, x), of real numbers, where the pair (θ, x) and $(\theta + n\,2\pi,\ \pm x)$ are identified, with the plus sign if n is an even integer, minus sign if an odd integer. Let M consist of numbers θ, with θ and $\theta + n\,2\pi$ (n an integer) identified. Thus, $M = S^1$ (setting $\tan\theta = y_2/y_1$ to obtain the usual representation of S^1). Set $\pi(\theta, x) = \theta$. The fibre over a point θ of M consists of all pairs, (θ, x), with vector space structure $(\theta, x) + c(\theta, y) = (\theta, x + cy)$. This is a vector bundle. Of course, the bundle space B is the Mobious strip.

Why bother with these vector bundles and "local product structures" instead of just considering products as in Example 38? Vector bundles have two nice properties that products do not. Firstly, Example 40 shows that vector bundles can have interesting global structure not possible for a product. Most vector bundles in practice are not products. Secondly, and more important, vector bundles have certain of the structure of a product "washed out". In the product, $M \times V^k$, one can say that two points, (p, y) and (q, w), lie "at the same places on their respective fibres" if $v = w$ (elements of V^k). In other words, there are natural isomorphisms between all the fibres. This is structure which is not present in a general vector bundle. In general, one cannot compare points on different fibres. This is what is wanted for applications.

Exercise 97. Define an isomorphism of vector bundles.

Exercise 98. Prove that B is connected if and only if M is.

Exercise 99. Prove that every vector bundle with base space $M = R^1$ is a product bundle.

Exercise 100. Prove that S^2 is not the bundle space for any vector bundle.

Exercise 101. Consider two vector bundles with the same base space M, and with fibres of dimension k and k'. By taking the direct sum (of vector spaces) within each fibre, obtain a third vector bundle with base space M, and fibres of dimension $(k + k')$.

17. The Tensor Bundles

With any manifold M there is associated a collection of vector bundles with base space M, and with fibre over a point p of M consisting of tensors in M at p. These are the bundles which are particularly interesting in differential geometry. We now introduce them.

Let M be an n−dimensional manifold. Denote by B the set consisting of all pairs (p, ξ^a), where p is a point of M and ξ^a is a contravariant vector in M at p. Define a mapping π from the set B to the manifold M as follows: $\pi(p, \xi^a) = p$. For each point p of M, introduce the obvious n−dimensional vector space structure on $\pi^{-1}[p]$, i.e., $(p, \xi^a) + c(p, \eta^a) = (p, \xi^a + c\eta^a)$. Our plan is to make (B, M, π) into a vector bundle. What must be done is to introduce a manifold structure on B, show that π is then a smooth mapping, and, finally, show that B is, locally, a product. These three things are all done at once. Let (U, φ) be a chart on M, so each point of U is labeled by n coordinates, $x^1, \ldots, x^n$. A contravariant vector in M at this point can be labeled by its n components with respect to this chart, $\tilde{\xi}^1, \ldots, \tilde{\xi}^n$. Thus, each point of the subset $\pi^{-1}[U]$ of B is labeled by $2n$ numbers, $(x^1, \ldots, x^n,\ \tilde{\xi}^1 \ldots, \tilde{\xi}^n)$. This will be a chart on B. These charts are clearly compatible, so we are led to a manifold structure on B. Furthermore, since π, applied to the point of B with coordinates $(x^1, \ldots, x^n,\ \tilde{\xi}^1, \ldots, \tilde{\xi}^n)$, is the point of M with coordinates $(x^1, \ldots, x^n)$, π is certainly smooth. Finally, it is also clear that B is locally a product. Let V^n be labeled by n−tuples, $(\tilde{\xi}^1, \ldots \tilde{\xi}^n)$, and let ψ, applied to the point of $U \times V^n$ represented by $(p; \tilde{\xi}^1 \ldots, \tilde{\xi}^n)$. That this ψ has all the required properties is, once again, clear. In words, a chart on M, since contravariant vectors can then be expressed in terms of components, induces a natural chart on B.

Thus, any manifold M defines a vector bundle in which the base space is M, the bundle space is the collection of contravariant vectors at points of M, the fibre over a point p of M is the vector space of contravariant vectors at p, and the projection takes a contravariant vector at a point of M to the point at which that vector is. This vector bundle is called the *tangent bundle* of M, TM. (Note that the bundle structure is just right here. It is meaningless to say that a contravariant vector at one point of M is "the same" as a contravariant vector at another point of M. The notation of a vector bundle, conveniently, also does not allow such a comparison.)

Similarly, the vector bundle with base space M, fibre over the point p of M the n–dimensional vector space of covariant vectors at p, bundle space the collection of all covariant vectors at points of M, etc. is called the *cotangent bundle* of M, CM.

More generally, consider tensors with a given index structure, e.g., $\alpha^{ac}{}_{rq}{}^{d}$, at points of M. Using exactly the same argument as above, we have a vector bundle with base space M, and fibre over the point p of M tensors in M at p with precisely this index structure. (In this example, $k = n^5$.) The vector space structure in each fibre is, of course, the vector space structure on tensors at a fixed point p of M. These are called the *tensor bundles* of M. (They are, of course, still vector bundles in the sense of Sect. 16.)

Example 41. Let M be a manifold. Let B be the collection of pairs $(p, \alpha^{ac}{}_{rq}{}^{d})$, where $\alpha^{ac}{}_{rq}{}^{d}$ is a tensor at p satisfying $\alpha^{bc}{}_{nq}{}^{d} = \alpha^{a[c}{}_{nq}{}^{d]}$ (or any other linear condition on α). Then, as above, we obtain a vector bundle with base space M.

Example 42. Let M be a manifold. Let B be the collection of triples, $(p,\beta^{a}{}_{b},\gamma^{de}{}_{f})$, where $\beta^{a}{}_{b}$, and $\gamma^{de}{}_{f}$ are tensors at p. Set $\pi(p,\beta^{a}{}_{b},\gamma^{de}{}_{f}) = p$. Define addition in the fibres by $(p,\beta^{a}{}_{b},\gamma^{de}{}_{f}) + c(p,\tau^{a}{}_{b},\mu^{de}{}_{f}) = (p,\beta^{a}{}_{b} + c\tau^{a}{}_{b},\gamma^{de}{}_{f} + c\mu^{de}{}_{f})$.

Then, as above we have a vector bundle with base space M.

Exercise 102. Why is there no vector bundle with "derivative operators" replacing tensors?. Why is there no vector bundle with "metrics" replacing tensor?

Exercise 103. Prove that all the tensor bundles of R^n are products.

Exercise 104. Using the fact that every vector field on S^2 vanishes at some point, prove that the tangent bundle of S^2 is not a product.

Exercise 105. Suppose we have a diffeomorphism between M and M'. Find a diffeomorphism between the bundle spaces of their cotangent bundles.

Exercise 106. Find a natural diffeomorphism between the bundle space of the tangent bundle of the product of two manifold and the product of the bundle spaces of the tangent bundles of the two manifolds.

Exercise 107. Why is R^7 not the bundle space of the cotangent bundle of any manifold?

Exercise 108. Show that a metric on M defines a natural diffeomorphism between the bundle space of the tangent bundle of M and the bundle space of the cotangent bundle of M. (Raising and lowering.)

Example 43. In classical mechanics, the configuration space of a system is a manifold M. The Lagrangian is a function on the tangent bundle of M. The cotangent bundle of M is called phase space, and the Hamiltonian is a function on the cotangent bundle.

Example 44. Let M be a manifold. There is a natural covariant vector field on the cotangent bundle of M. Let P be a point of CM, so P consists of a pair (p, μ_a), where μ_a is a covariant vector in M at the point $p = \pi(P)$ of M. A covariant vector in CT at P assigns, to each contravariant vector in CT at P, a real number. Let Λ be a contravariant vector in CT at P, and consider the number $\mu_a(\lambda^a)$ where $\lambda^a = \vec{\pi}_P \Lambda$ is a contravariant vector in M at p. This assignment defines a covariant vector at each point P of CM, and hence a covariant vector field (Exercise 109. Prove smooth.) on CM. In classical mechanics, the exterior derivative of this covariant vector field is the symplectic structure on phase space.

Let M be a manifold, and let B be one of the tensor bundles over M. A mapping which takes each point p of M to a point of the fibre over p certainly assigns a tensor of the appropriate type to each point of M, and conversely. In other words, there is a natural, one-to-one correspondence between (not necessarily smooth) tensor fields on M and (not necessarily smooth) mappings $\Delta : M \to B$ which satisfy $\pi \cdot \Delta =$ the identity diffeomorphism on M. We prove:

Theorem 15. The tensor field is smooth if and only if the mapping $\Delta : M \to B$ is smooth.

Proof: We do the case for the tangent bundle, all others being identical. Let (U, φ) be a chart on M. Then $\Delta : M \to B$ is represented by n functions of n variables: $\Delta(x^1, \ldots, x^n) = (x^1, \ldots, x^n, \xi^1(x), \ldots, \xi^n(x))$. Clearly, Δ is smooth if and only if, for every such chart, the functions $\xi^1(x), \ldots, \xi^n(x)$ are C^∞. But a contravariant vector field on M is smooth if and only if its components in every chart are C^∞ functions of the coordinates.

Let (B, M, π) be a vector bundle. A smooth mapping $\Delta : M \to B$ such that $\pi \cdot \Delta =$ identity on M is called a smooth *cross section*. Thus, for the tensor bundles, the cross sections are precisely the smooth tensor fields. We could have just as well have defined a smooth tensor field as one obtained as a smooth cross section, so smoothness of tensor fields would be obtained from smoothness of maps.

Exercise 110. prove that every vector bundle possesses a cross section.

The fibre over p in the tangent bundle (resp. cotangent bundle) is called the *tangent space* (resp. *cotangent space* of p).

18. Smooth Mappings: Action on Tensor Bundles

Let $\psi : M \to M'$ be smooth. Recall, from Sect. 15, that this ψ induces a mapping $\vec{\psi}_p$ from contravariant tensor in M at p to contravariant tensors in M' at $p' = \psi(p)$, a mapping $\underset{\leftarrow}{\psi}_p$ from covariant tensors in M' at p' to covariant tensors in M at p, and a mapping $\underset{\leftarrow}{\psi}$ from covariant tensor fields on M' to covariant tensor fields on M'. This discrimination against contravariant tensors is corrected, to a certain extent, when we ask what mappings ψ induces on the corresponding tensor bundles. We now ask this.

Let $\psi : M \to M'$ be smooth. Then, for each point p of M, $\underset{\leftarrow}{\psi}_p$ is a linear mapping from the cotangent space of $p' = \psi(p)$ to the cotangent space of p. Does this $\underset{\leftarrow}{\psi}_p$ induce a mapping from the cotangent bundle of M', CM', to CM? No. For one thing, certain points of M' may not even be in the range of ψ, so the fibres over such points would not have anywhere to be mapped to. Furthermore, two distinct points, p and q, of M might be mapped, by ψ, to the same point of M', whence there would be ambiguity as to where the cotangent space of this point of M' should be mapped to.

Of course, it works the other way around. Let p be a point of M, ξ^a a vector in M at p, so (p, ξ^a) is a point of TM. Then $(\psi(p), \vec{\psi}_p(\xi^a))$ is a point of TM', for $\vec{\psi}_p\ \xi^a$ is a contravariant vector in M' at the point $\psi(p)$ of M'. Hence, ψ induces a mapping $\overset{\leadsto}{\psi} : TM \to TM'$. We prove that this mapping is smooth. Let (U, φ) be a chart on M, and (U', φ') a chart on M'. Then $x^1, \dots, x^n$ are coordinates in U, while $x'^1, \dots, x'^{n'}$ are coordinates in U'. In terms of these charts, the action of ψ is represented by n', C^∞ functions of n variables, $x'^1(x^1, \dots, x^n), \dots, x'^{n'}(x^1, \dots, x^n)$. A point of TM in $\pi^{-1}[U]$ is represented by a $2n$-tuple of numbers $(x^1, \dots, x^n, \tilde{\xi}^1, \dots, \tilde{\xi}^n)$, and similarly for M'. In terms of these charts on TM and TM', the action of $\overset{\leadsto}{\psi}$ is as follows: $\overset{\leadsto}{\psi}(x^1, \dots, x^n, \tilde{\xi}^1, \dots, \tilde{\xi}^n) = (x'^1(x), \dots, x'^{n'}(x), \Sigma_i \tilde{\xi}^i \frac{\partial x'^j}{\partial x^i}|_x \dots, \Sigma_i \tilde{\xi}^i \frac{\partial x'^{n'}}{\partial x^i}|_x)$. That is, we have $2n'$ functions of $2n$ variable. Since these functions are C^∞, the mapping $\overset{\rightsquigarrow}{\psi}$ is smooth.

The general situation is now clear. $\psi : M \to M'$ induces a smooth map, $\overset{\rightsquigarrow}{\psi}$, from a tensor bundle over M to a tensor bundle over M', provided the tensor bundle over M consists of tensors all of whose indices are contravariant. No natural mappings are induced on the mixed bundles or covariant tensor bundles.

Finally, we note that $\psi \cdot \pi = \pi' \cdot \overset{\rightsquigarrow}{\psi}$, and equation which states that the fibre over a point p of M is mapped, by $\overset{\rightsquigarrow}{\psi}$ to the fibre over $\psi(p)$ of M'.

The second half of the table on p. 68 can now be written in a more symmetrical form:

$\psi : M \to M'$	contra- variant tensor fields	co- variant tensor fields	mixed tensor fields	contra- variant tensor bun- dles	co- variant tensor bun- dles	mixed tensor bun- dles
From M to M'	—	—	—	$\overset{\rightsquigarrow}{\psi}$	—	—
From M' to M	—	$\underset{\leftarrow}{\psi}$	—	—	—	—

Of course, if $\psi : M \to M'$ is a diffeomorphism, then ψ induces a diffeomorphism between all the tensor bundles of the same index type, this diffeomorphism preserving all structure (i.e., their fibres, the vector-space structure, etc.).

Exercise111. Let $\psi : M \to M'$ and $\Lambda : M' \to M''$ be smooth. Prove that $\overset{\rightsquigarrow}{\Lambda} \cdot \overset{\rightsquigarrow}{\psi} = (\overset{\rightsquigarrow}{\Lambda \cdot \psi})$.

Exercise 112. Show that a metric on M induces a natural function $(g_{ab}\xi^a\xi^b)$ on TM. Prove that this function is smooth.

Exercise 113. In sect. 15, we argued that scalar fields should be regarded as covariant tensor fields of rank zero, for $\underset{\leftarrow}{\psi}$ acts on them. Introduce the vector bundle for which scalar fields are the cross sections, and show that $\overset{\rightsquigarrow}{\psi}$ acts on this bundle. Hence, scalar fields should also be regarded as contravariant tensor fields of rank zero.

Exercise 114. Let ψ be smooth from M to M'. Find an example in which ψ is onto but $\overset{\rightsquigarrow}{\psi}$ is not. In which ψ is one-to-one, but $\overset{\rightsquigarrow}{\psi}$ is not.

19. Curves

Let M be a manifold. By a (smooth) *curve* in M we understand a smooth mapping $\gamma : I \to M$ from some open interval $I = (a, b)$ a possibly $-\infty$ or b $+\infty$ of the real line to M. (Note that I, as an open subset of R^1, inherits a manifold structure.) If t is a real number in the interval I we write $\gamma(t)$ for the image of t under γ. Note that, by this definition, a choice of parameterization is part of a curve.

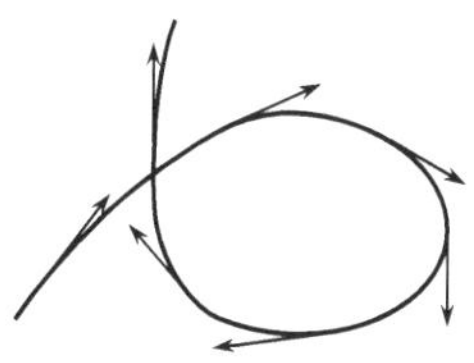

We define the tangent vector to a curve. Since the interval I is a subset of the reals R, the manifold I has an obvious natural chart on it. Denote by i^a the contravariant vector field on I whose component, with respect to this chart, is (1). Thus, if $f(t)$ (t in I) is a smooth function on I, then $i^a\ D(f)_a$ is the function df/dt on I. For each point t of I, $\vec{\gamma}_t(i^a)$ is, therefore, a contravariant vector in M at the point $\gamma(t)$ of M. This vector is called the *tangent vector* to the curve at t. Of course, the curve may "cross itself", i.e., we may have $\gamma(t_1) = \gamma(t_2)$ for $t_1 \neq t_2$. The tangent vectors at t_1 and t_2 may be different for such a crossing point, i.e., we may have $\vec{\gamma}_{t_1}\ (i^a) = \vec{\gamma}_{t_1}\ (i^a)$.

Example 45. In terms of a chart, a curve is represented by n functions of one variable, $x^1(t), \ldots, x^n(t)$. The tangent vector to this curve at t_0 is the vector at the point with coordinates $x^1(t_0), \ldots, x^n(t_o)$ with components $(dx^1/dt|_{t_0}, \ldots, dx^n/dt|_{t_0})$.

Example 46. Let p be a point of the manifold M, and let $\gamma : I \to M$ be the constant curve, $\gamma(t) = p$, which just remains at p. We show that its tangent vector is zero. If h is any smooth function on M, $\underleftarrow{\gamma}, h$ is a constant function on I, namely the one with value $h(p)$. So, $i^a D(\underleftarrow{\gamma} h)_a = 0$. So, $\vec{\gamma}_t\ i^a = 0$.

Example 47. Let f be a smooth function on M, and consider $f : M \to R$ as a smooth mapping of manifolds. Denote by i_a the covariant vector field on the manifold R whose component, in the natural chart, is (1). Then $\underleftarrow{f}(i^a) = D(f)_a$.

We are now in a position to give an intuitive geometrical interpretation of a contravariant vector at a point of a manifold M as an "infinitesimal displacement". Let p, be a point of M. In order to discuss infinitesimal displacements from p, we need a family of points of M which approach p. A curve does the job.

So, let $\gamma : I \to M$ be a curve, say, with $\gamma(t_0) = p$. Let ξ^a be the tangent vector to γ at t_0. Then, one should interpret "$\xi^a dt$ as the infinitesimal displacement vector from the point $p = \gamma(t_0)$ of M to the point $\gamma(t_0 + dt)$ of M." One justification for this interpretation would be the following. Set $\tilde{f} = \overleftarrow{\gamma} f$, where f is any smooth function on M. Thus, $\tilde{f}(t)$ is just a smooth function on I. Now,

$$\frac{d}{df}\tilde{f}(t)\bigg|_{t_0} = i^a D(\tilde{f})_a|_{t_0} = i^a D(\overleftarrow{\gamma}\tilde{f})_a|_{t_0} = i^a \overleftarrow{\gamma} D(f)_a|_{t_0}$$
$$= (\vec{\gamma}_p i^a) D(f)_a|_p = \xi^a D(f)_a|_p$$

Whereas contravariant vectors have this interpretation as "infinitesimal displacements," no such interpretation is available for covariant vectors.

Example 48. The interpretation discussed above can also be obtained, more directly, from Example 45.

Thus, a metric, g_{ab}, on M associates, with each contravariant vector ξ^a at a point, a number, $R_{ab}\ \xi^a \xi^b$. The metric can therefore be thought of as describing a distance between infinitesimally nearby points (at least, if g_{ab} is positive-definite). Since a knowledge of $R_{ab}\ \xi^a \xi^b$ for every contravariant ξ^a at a point determines the metric g_{ab} uniquely, a metric represents nothing more than the information of these distances between infinitesimally nearby points.

Example 49. Let g_{ab} be a positive-definite metric on M. We define the *length* of the curve $\gamma : I \to M$, where $I = (a, b)$. For each t in I, set $\lambda(t) = g_{ab}\ \xi^a \xi^b$, where ξ^a is the tangent vector to γ at t. Thus, $\lambda(t)$ is a function of one variable. Then (length of γ)= $\int_a^b [\lambda(t)]^{1/2} dt$. (Exercise 115. Verify that this length is independent of the parameterization. That is, let J be another open interval, and let $\mu : J \to I$ be a diffeomorphism. Then $\gamma \cdot \mu : J \to M$ is a curve in M. (It is practically the same curve. It hits the same points of M, but these are parameterized in a different way.) Verify that the length of $\gamma \cdot \mu$ is the same as that of γ.)

Finally, note that, if $\psi : M \to M'$ is a smooth mapping, and $\gamma : I \to M$ is a curve in M, then $psi \cdot \gamma : I \to M'$ is a curve in M'. Under smooth mapping, curves are sent the same way as contravariant tensors.

20. Integral Curves

Of course, a manifold M possesses a great number of curves. When a vector field is specified on M, then a certain collection of curves, defined by this vector field, are often of particular interest.

Let ξ^a be a contravariant vector field on M. A curve $\gamma : I \to M$ in M is said to be an *integral curve* of ξ^a if, for each point t of I, the tangent vector to $\gamma(t)$ at t (a contravariant vector in M at the point $\gamma(t)$ of M) coincides with the contravariant vector field ξ^a evaluated at $\gamma(t)$. Intuitively, an integral curve of a vector field "runs along in the direction of the vector field, moving quickly (i.e., covering a lot of M with each increment of t) where ξ^a is large, and slowly where ξ^a is small". If the interval I contains the origin 0 of R, then the point $\gamma(0)$ of M is called the *initial value* of the integral curve.

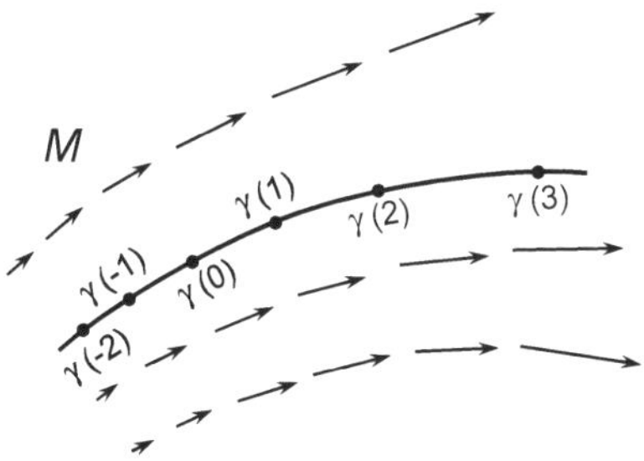

Let $\gamma : I \to M$ be an integral of ξ^a, with $I = (a, b)$. Let c be a real number, and set $I' = (a + c, b + c)$, and $\gamma'(t) = \gamma(t - c)$ for t in I'. Then $\gamma' : I' \to M$ is also an integral curve. (Proof: The diffeomorphism from I to I', which sends $t+c$ clearly takes the contravariant vector field i^a on I to i^a on I'.) If 0 is in both I and I', then the initial value of γ', $\gamma'(0)$. is the point $\gamma(-c)$ of M. Thus, all we have done is reparameterized the curve by adding a constant (namely c) to the perimeter. The consequence is to shift the initial value of $\gamma(t)$ to another point of the curve. It is clear that, given an integral curve, and a point p (in M) on that curve, one can, by reparameterizing in this way, obtain a new curve with initial value p.

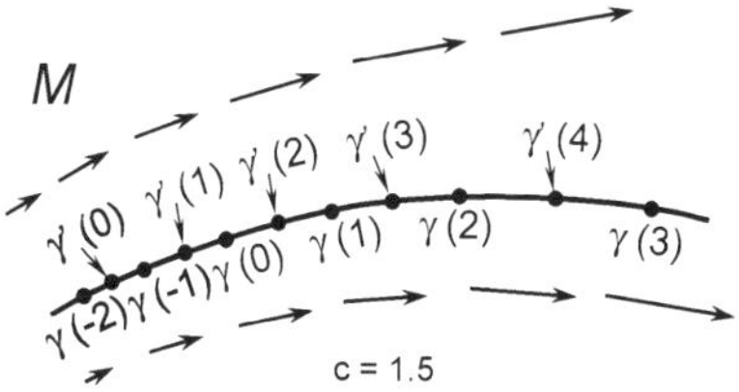

We shall call the operation described above *shifting the initial value*.

A fundamental property of integral curves is the following:

Theorem 16. Let ξ^a be a contravariant vector field on the manifold M, and let p be a point of M. Then there exists an integral curve of ξ^a, $\gamma : I \to M$, with initial value p, and with the following property: if $\gamma' : I' \to M$ is another integral curve of ξ^a, with initial value p, then $I' \subset I$, and for all t in I', $\gamma'(t) = \gamma(t)$

The proof of Theorem 16, although not very difficult, is rather technical and long. We omit it. A proof can be found in most textbooks on ordinary differential equations. It is immediate that the curve $\gamma : I \to M$ whose existence is guaranteed by Theorem 16 is unique, (Proof: If $\gamma : I' \to M$ were another, then we would have $I' \subset I$ and $I \subset I'$, so $I = I'$, and also $\gamma'(t) = \gamma(t)$ for t in I'. In other words, we would have the fact that γ' and γ are the same curve.) The unique curve obtained in Theorem 16 will be called the *maximal* integral curve (of ξ^a) with initial value p.

Theorem 16 is the basic existence and uniqueness theorem for ordinary differential equations. The following example illustrates this remark:

Example 50. Consider the pair of coupled ordinary differential equations

$$f' = \cos(\tau f) - g'\, e^g \tau^3 \qquad g'' = \frac{2f^2 g g'\, \tau}{f^2 + (g')^2} \tag{91}$$

for function $f\tau$) and $g(\tau)$, where a prime denote $d/d\tau$. Denote by M the four-dimensional manifold consists of R^4 with the (closed) region $(x^1)+(x^3)^2 = 0$ removed. We have a natural chart on M. Let ξ^a be the vector field on M whose components, with respect to this chart, are

$$\xi^1 = \cos(x^4 x^1) - x^3\, e^{(x^2)}(x^4)^3 \qquad \xi^3 = 2(x^1)^2\, x^2\, x^3\, x^4\, [(x^1)^2 + (x^3)^2]^{-1}$$
$$\xi^2 = x^3 \qquad \xi^4 = 1$$

This ξ^a is clearly smooth on M. Let p denote the point of M with coordinates $(a, b, c, 0)$. Then, by Theorem 16, there exists a unique maximal integral curve, $x^1(t), \ldots, x^4(t)$, of ξ^a, with initial value p. That is, by Example 45, we have

$$\frac{d}{dt}x^1 = \cos(x^4\, x^1) - x^3\, e^{(x^2)}\, (x^4)^3 \qquad \frac{d}{dt}x^3 = 2(x^1)^2\, x^2\, x^3 x^4 \left[(x^1)^2 + (x^3)^2\right]^{-1} \tag{92}$$

$$\frac{d}{dt}x^2 = x^3 \qquad \frac{d}{dt}x^4 = 1$$

with

$$x^1(0) = a \qquad x^3(0) = c$$
$$x^2(0) = b \qquad x^4(0) = 0$$

Clearly, $x^4(t) = t$. Then, from (92), the functions $f(\tau) = x^1(\tau)$ and $g(\tau) = x^2(\tau)$, satisfy the differential equations (91). (Note that $x^3(\tau) = g'(\tau)$.) This solution has initial values $f(0) = a$, $g(0) = b$, and $g'(0) = c$. In other words, Theorem 16 ensures that, given arbitrary values of f, g, and g' at $\tau = 0$, (with $(f(0))^2 + (g'(0))^2 = 0$), the differential equation (91) has precisely one solution extended maximally in τ.

A number of properties of integral curves follow immediately from Theorem 16. Let $\gamma : I \to M$ be a maximal integral curve, with $I = (a, b)$ (a may be $\mp\infty$, or $b + \infty$, or both). Let d be a number in the interval (a, b). Then, as we remarked before, $\gamma' : J \to M$, defined by $\gamma'(t) = \gamma(t + d)$ where $J = (a - d, b - d)$, is an integral curve. In fact, γ' is maximal. (Proof: If the t–interval J of γ' could be enlarged, then, by shifting the initial value, we would have an enlargement of the t–interval I of γ. But this would contradict the assumed maximality of γ.) Thus, shifting of the initial value, applied to a maximal integral curve, results in a maximal integral curve.

Integral curves cannot cross. That is, if $\gamma : I \to M$ and $\gamma' : I' \to M$ are maximal integral curves, and if $\gamma(t_1) = \gamma'(t_2) = p$ for some t_1 and t_2, then γ and γ' are obtainable from each other by shifting the initial value. Proof: By shifting initial values, we can have p the initial value of γ and of γ'. By Theorem 16, these curves are then identical. Hence, the original curves differ from each other only by shifting of the initial value.

As a final illustration of the use of Theorem 16, we remark that, if p is a point of M at which ξ^a vanishes, and if $\gamma : I \to M$ is an integral curve passing through p (i.e., if $\gamma(t) = p$ for some t), then γ is the constant curve: $\gamma(t) = p$ for all t. Proof: By shifting the initial value, we might as well take p as the initial value of γ. Let $\tilde{\gamma} : R \to M$ be the constant curve, $\tilde{\gamma}(t) = p$ for all t. Then, by Example 46, $\tilde{\gamma}$ is an integral curve of ξ^a. Since the interval is the whole real line, this integral curve must be maximal. Hence, by Theorem 16, γ is the constant curve remaining at p.

Numerous other – intuitively clear – properties of integral curves follow directly and similarly from Theorem 16.

We next discuss the dependence of the maximal integral curves on the initial value. For this purpose, it is convenient to introduce the following definition. A contravariant vector field ξ^a on M is said to be *compete* if every maximal integral curve has $I = (-\infty, +\infty)$. Intuitively, a complete vector field has the property that ξ^a becomes "small" as one approaches the "edge" of M, so only "in the limit $t \to \infty$ or $t \to -\infty$ do integral curves approach the edge". (Note: It is more conventional to define completeness of a vector field by requiring only that the intervals have the form $(a, +\infty)$, where a may be finite or $-\infty$,)

Exercise 116. Let ξ^a be a complete on M, and let C be a closed subset of M. Prove that ξ^a is complete on $M - C$ and if and only if every integral curve with initial value in $M - C$ remains in $M - C$.

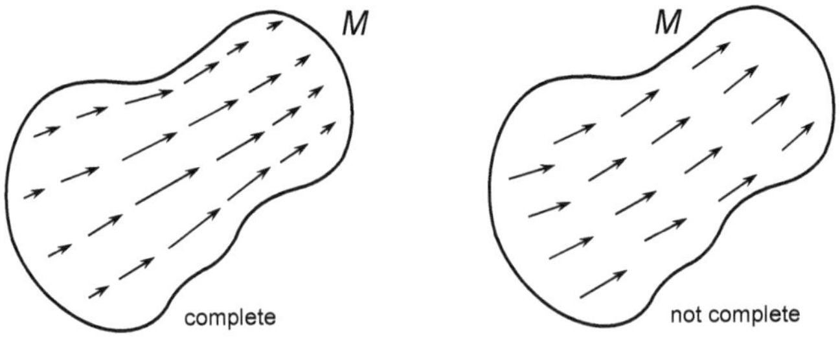

Exercise 117. Let ξ^a be complete on M, and let p be a point of M. Prove that ξ^a is complete on $M - p$ if and only if ξ^a vanishes at p.

Exercise 118. Find an example of two complete vector fields whose sum is not complete; of two vector fields neither of which is complete, but whose sum is complete.

Exercise 119. Prove that ξ^a is complete if and only if $-\xi^a$ is complete.

Let ξ^a be a complete vector field on M. We define a mapping $\Gamma : R \times M \to M$ as follows. If t is a number, p a point of M, $\Gamma(t, p) = \gamma(t)$, where γ is the maximal integral curve with initial value p. (Exercise 120. Why do we need completeness for this?) The second fundamental property of integral curves is the following:

Theorem 17. Let ξ^a be a complete contravariant vector field on M. Then $\Gamma : R \times M \to M$ is smooth.

Again, we omit the proof. See most textbooks on ordinary differential equations. In terms of differential equations, Theorem 17, states that the solution of s differential equation depends smoothly on the initial value.

Write $\Gamma_t(p)$ for $\Gamma(t, p)$. Then, fixed t, $\Gamma_t : M \to M$ is a mapping from M to M. Intuitively, the action of Γ_t is as follows: "move each point p of M a parameter-distance t along the integral curve through p". Thus, points of M where ξ^a is "large" are "moved a great deal" by Γ_t, points where ξ^a is small are moved less far, and points of M at which ξ^a vanishes are left invariant by Γ_t.

Exercise 121. Prove that, if ξ^a vanishes at p, then $\Gamma_t(p) = p$ for all t.

We prove that Γ_t is smooth for each t. Fix t, and let $\Lambda; M \to R \times M$ be defined by $\Lambda(p) = (t, p)$. Then Λ is smooth. Furthermore, $\Gamma_t = \Gamma \cdot \Lambda$. By Theorem 17, Γ_t is smooth. Note, furthermore that, by shifting the origin, Γ_{-t} is the inverse of Γ_t (i.e., $\Gamma_{-t} \cdot \Gamma_t = \Gamma \cdot \Gamma_{-t}$ = identity mapping on M). Thus, Γ_t is one-to-one and onto. It's inverse, Γ_{-t}, is also smooth (for Γ_ω is smooth for all ω, in particular, for $\omega = -t$). Thus,

Theorem 18. For each number t, $\Gamma_t : M \to M$ is a diffeomorphism. Furthermore, $\Gamma_t \cdot \Gamma_\omega = \Gamma_{(t+\omega)}$.

These remarks further strengthen the interpretation of contravariant vectors as "infinitesimal displacements". The mapping Γ_t represents the corresponding "finite displacement" obtained by "integrating the infinitesimal displacement defined, at each point, by ξ^a". So, "Γ_{dt} represents the corresponding infinitesimal displacement".

Exercise 122. Consider the vector field in R^2 with components $\xi^1 = x^2, \xi^2 = -x^1$. Prove that this field is complete. Find an explicit formula for Γ_t. Check Theorem 17 and 18 explicitly.

Example 51. In classical mechanics, the vector field $\xi^a = F^{ab}\ D(H)_t$, in phase space, where F^{ab} is the inverse of the symplectic 2−form at each point (i.e., $F^{ab}\ F_{cb} = \delta^a{}_c$) and H is the Hamiltonian function, is called the Hamiltonian vector field. The integral curves of this (normally complete) vector field describe the evolution of the system in time.

Example 52. In fluid mechanics, the velocity field of the fluid is a complete vector field in Euclidean 3−space. Then Γ_t maps the locations of fluid elements at time zero to their locations at time t. A vector field is sometimes called a flow.

Exercise 123. Find a nonzero vector field such that Γ_t = identity for some t.

Exercise 133. Find an example of a diffeomorphism from a manifold to itself which is not a Γ_t for any vector field or any t.

Exercise 134. Prove Theorems 16 and 17 in the case when M is one-dimensional.

21. The Lie derivative: Geometrical Interpretation

We have already discussed the Lie derivative from two different points of view: the algebraic approach (Sect. 7), and the concomitant approach (Sect. 11). The most intuitive, and often the most useful, approach is the one we now introduce.

Let ξ^a be a complete vector field on M. (Completeness merely serves to simplify the discussion here. It makes the range and domain of Γ_t be M. With slightly more effort, but with little increase in content, one could drop the completeness assumption in this section.) Let $\alpha^{\cdots}{}_{\cdots}$ be a tensor field on M. Now, for each number t, Γ_{-t} is a diffeomorphism on M. Thus, Γ_{-t} takes the tensor field $\alpha^{a\ldots c}{}_{b\ldots d}$ on M to some other tensor field, $\alpha^{a\ldots c}{}_{b\ldots d}(t)$, on M. That is to say, we have for each t, a tensor field $\alpha^{a\ldots c}{}_{b\ldots d}(t)$ on M. That is to say, we have, for each t, a tensor field $\alpha^{a\ldots c}{}_{b\ldots d}(t)$ on M. One says that $\alpha^{a\ldots c}{}_{b\ldots d}(t)$ results from $\alpha^{a\ldots c}{}_{b\ldots d}(0)$ (our original tensor field) by dragging along the vector field ξ^a.

We propose to prove that

$$\frac{d}{dt}\alpha^{a\ldots c}{}_{b\ldots d}(t)\bigg|_{t=0} = \mathscr{L}_\xi \alpha^{a\ldots c}{}_{b\ldots d} \tag{93}$$

where the left side has the following meaning. Fix a point p of M. Then, for each t, $\alpha^{a\ldots c}{}_{b\ldots d}(t)$ is a tensor at p. Then $\frac{d}{dt}\alpha^{a\ldots c}{}_{b\ldots d}(t)|_0$ is the tensor field whose value at p is

$$\lim_{\Delta t\to 0}\frac{1}{\Delta t}\left[\alpha^{a\ldots c}{}_{b\ldots d}(\Delta t) - \alpha^{a\ldots c}{}_{b\ldots d}(0)\right]$$

Fix a point p of M, and let $\gamma : R \to M$ be the (maximal) integral curve with initial value p. We first establish (93) for a scalar field α. Let $\alpha(t) = \underleftarrow{\Gamma_t}\alpha$, and let $f(t)$ be the scalar field $\alpha(t)$ evaluated at p, so $f(t)$ is one real function of the variable. Then

$$\frac{d}{dt}f\Big|_0 = i^a\ D(f)_a|_0 = i^a\ D(\underset{\leftarrow}{\Gamma_t}\ \alpha)_a|_0 = i^a\underset{\leftarrow}{\Gamma_t}\ D(\alpha)_a|_0$$
$$= \left((\vec{\Gamma}_t)_p i^a)\right) D(\alpha)_a\Big|_p = \xi^a D(\alpha)_a\Big|_p = (Z)_\xi \alpha\Big|_p$$

Since this equation holds for every point p, Eqn. (93) holds for scalar fields. Next, let $\omega_{a...c}$ be a form. Then, since exterior differentiation is preserved by a diffeomorphism, we have, for every t,

$$\Gamma_t D(\omega)_{ma...c} = D\ (\Gamma_t\omega)_{ma...c} \tag{94}$$

Taking d/dt of this equation, and evaluating at $t = 0$,

$$\frac{d}{dt}\ D\Big(\omega(t)\Big)_{ma...c} = D\Big(\frac{d}{dt}\omega(t)\Big)_{ma...c} \tag{95}$$

Let α be a scalar vector. Then

$$\frac{d}{dt}D(\alpha)_a\Big|_0 = D(\frac{d}{dt}\alpha)_a\Big|_0 = D(\mathscr{L}_\xi\ \alpha)_a = \mathscr{L}_\xi\ D(\alpha)_a$$

where we have used (95) and the fact that the Lie derivative and exterior derivative commute. Thus, (93) holds for a gradient. If α and β are scalar field, therefore,

$$\frac{d}{dt}\Big(\beta(\alpha)_a\Big)\Big|_0 = \Big(\frac{d}{dt}\beta\Big)\Big|_0 D(\alpha)_a + \beta\frac{d}{dt}D(\alpha)_a\Big|_0 = (\mathscr{L}_\xi\beta)\ D(\alpha)_a + \beta\mathscr{L}_\xi D(\alpha)_a$$
$$= \mathscr{L}_\xi\Big(\beta D(\alpha)_a\Big)$$

But, since every covariant vector field can, locally, be expressed as a sum of products of scalar fields and gradients, Eqn. (93) holds for arbitrary covariant vector fields. Next, let η^a be a contravariant vector field, μ_a a covariant vector field. Then

$$\mu_a\frac{d}{dt}\eta^a\Big|_0 - \frac{d}{dt}(\mu_a\eta^a)\Big|_0 - \eta^a\frac{d}{dt}\mu_a\Big|_0 = \mathscr{L}_\xi(\mu_a\eta^a) - \eta^a\,\mathscr{L}_\xi\mu_a$$
$$= \mu_a\mathscr{L}_\xi\eta^a$$

Since μ_a is arbitrary, (93) holds for $\alpha^{a...c}{}_{b...d}$ a contravariant vector field. Finally, for an arbitrary tensor field, we proceed as usual:

$$\mu_a\dots\nu_c\ \eta^b\dots\tau^d\ \frac{d}{dt}\alpha^{a...c}{}_{b...d}\Big|_0 = \frac{d}{dt}[\alpha^{a...c}{}_{b...d}\mu_a\dots\nu_c\eta^b\dots\tau^d]|_0$$
$$-\ \alpha^{a...c}{}_{b...d}\Big[(\frac{d}{dt}\mu_a)\dots\nu_c\eta^b\dots\tau^d + \dots + \mu_a\dots nu_c\eta^b\dots\frac{d}{dt}\tau^d\Big]\Big|_0$$
$$= \mu_a\dots\nu_c\eta^b\dots\tau^d\,\mathscr{L}_\xi\alpha^{a...c}{}_{b...d}$$

nothing that, since $\mu_a, \ldots, \nu_c, \eta^b, \ldots, \tau^d$ are arbitrary, (93) is proven.

Thus, ξ^a defines a one-parameter family of motions on M, each of which is a diffeomorphism. Each diffeomorphism carries tensor fields on M to fields on M. The "rate of change" of a tensor field under these motions is the Lie derivative of that field in the ξ^a–direction.

The fact that Lie derivatives habitually commute with other operations (e.g., exterior derivatives, contraction) should now be clear. Similarly, the Leibnitz rule for Lie derivatives, etc. is immediate from (93).

Exercise135. Let ξ^a be a vector field, and ∇_a derivative operator on M. Then, for each t Γ_{-t} takes ∇_a to another derivative operator, $\nabla_a(t)$, on M. These two operators are related by some tensor field, $\gamma^m{}_{ab}(t)$, on M. Prove that

$$\frac{d}{dt}\gamma^m{}_{ab}\bigg|_0 = -\nabla_a \nabla_b \xi^m + R_{sab}{}^m \xi^s$$

The remarks surrounding Eqn. (66) now make sense.

Exercise136. Do Exercise 82 in one sentence.

Example 53. In general relativity, because of the interpretation above, Killing vectors represent symmetries in space-time.

Exercise 137. Let $\Gamma : R \times M \rightarrow M$ be smooth, and suppose that, for each point p of M, $\Gamma(t, \Gamma(w, p)) = \Gamma(t + w, p)$. Find a contravariant vector field on M such that it generates Γ (in the sense of Theorem 17).

Exercise 138. Prove Eqn. (31) directly from (93).

Exercise 139. In Exercise 122, let g_{ab} be the metric with components $g_{11} = g_{22} = 1, g_{12} = g_{21} = 0$. Prove that $\mathcal{L}_\xi g_{ab} = 0$ in two ways, once directly (e.g., using concomitants), and once using (93).

22. Lie Groups

As an example and application of some of the preceding material, we obtain a few properties of Lie groups.

Roughly speaking, a Lie group is both a group and a manifold, where these two structures interact in a natural way. Let G be a set. Suppose that we are given a group structure on G, i.e., suppose we are given a mapping $\mathscr{C} : G \times G \to G$ (composition) and a mapping $\mathscr{J} : G \to G$ (inversion), subject to the usual axioms for a group. One normally writes $\alpha\beta$ instead of $\mathscr{C}(\alpha,\beta)$and $\alpha{-}1$ instead of $\mathscr{J}(\alpha)$. Suppose further that we are given a manifold structure on G, i.e., suppose we are given a collection of charts on G satisfying $M1 - M4$. We now have both structures on G: we have yet to require that they interact properly with each other. The most natural "interaction" available is the requirement that the mapping $\mathscr{C} : G \times G \to G$ and $\mathscr{J} : G \to G$ (which define the group structure) are smooth (as mappings of manifolds). Thus, a *Lie group* is a set G with both a group and a manifold structure, such that $\mathscr{C}$ and $\mathscr{J}$ are smooth.

Example 54. The Lorentz group, the Poincaré group, $SU(3)$ are all Lie groups.

Example 55. Let V be a k−dimensional vector space. Considering V as a manifold R^k (by choosing a basis, so each element of V is defined by its components, and so represents an element of R^k), and as an abelian group (under addition of vectors), V^k becomes a Lie group.

Fix an element α of the Lie group G, and let $\Delta_\alpha : G \to G$ be defined by $\Delta_\alpha(\beta) = \alpha\beta$, for any β in G. Then, evidently, we have $\Delta_\alpha \cdot \Delta_\gamma = \Delta_{\alpha\gamma}$, and Δ_e = identity mapping from G to G, where e is the identity element of the group G. In particular, for each α in G, $\Delta_\alpha \cdot \Delta_{\alpha^{-1}} = \Delta_{\alpha^{-1}} \cdot \Delta_\alpha$ = identity mapping from G to G. Thus, each Δ_α is a one-to-one onto mapping from G to G. We show that Δ_α, for fixed α, is a diffeomorphism. Note that the mapping $\lambda : G \to G \times G$ which sends β to (α,β) is smooth (injection of one factor into the product). But $\Delta_\alpha = \mathscr{C} \cdot \lambda$, and so Δ_α, as a composition of smooth maps, is smooth. Thus, since Δ_α is smooth, one-to-one, and onto, with smooth inverse, Δ_α is a diffeomorphism. Note, also that $\Delta_\alpha(e) = \alpha$.

A tensor field $\mu^{a...c}{}_{b...d}$ on G will be said to be *left invariant* if, for each

α in G, the diffeomorphism Δ_α takes the tensor field $\mu^{a\ldots c}{}_{b\ldots d}$ to itself. How does one obtain left invariant tensor fields on a Lie group? Let $\mu^{a\ldots c}{}_{b\ldots d}$ be any tensor in G at the point e (the identity). For each α in G, Δ_α is a diffeomorphism on G which takes e to α. Hence, Δ_α takes the tensor $\mu^{a\ldots c}{}_{b\ldots d}$ at e to some tensor at α. Repeating this for each point α of G we obtain a tensor at each point of G. That is, we obtain a tensor field on G. In fact, this tensor field is left invariant. This is clear, for, if α and β are two points of G, then Δ_β takes the tensor at e to the tensor at β, while $\Delta_{\alpha\beta} = \Delta_\alpha \cdot \Delta_{beta}$ takes the tensor at e to the tensor at $\alpha\beta$. Hence, Δ_α takes the tensor at β to the tensor at $\alpha\beta$. Since β is arbitrary, the e tensor field remains invariant under Δ_α. Since α is arbitrary, the tensor field is left invariant. We have:

Theorem 19. Given a tensor at e, there is precisely one left invariant tensor field on G which coincides with the given tensor at e.

It follows immediately that all the tensor bundles of a Lie group are product bundles. Thus, for example, we can conclude that S^2 cannot be the manifold of any Lie group (Exercise 104).

We define the Lie algebra of a Lie group. Intuitively, the Lie algebra is obtained by "commuting elements which differ infinitesimally from the identity". This remark suggests that we consider contravariant vectors at the identity e of G. Let ξ^a and η^a be contravariant vectors in G at e. Then, by Theorem 19, we have left invariant vector fields ξ^a and η^a on G. Clearly, $\mathcal{L}_\xi \eta^a = [\xi, \eta]^a$ is also left invariant. Evaluating at e, we obtain a contravariant vector in G at e. Thus, with two contravariant vectors at e, we associate a third, where this association is clearly linear. Thus, we have a tensor $C^m{}_{ab}$ at e such that $C^m{}_{ab}\xi^a\eta^b$ is the contravariant vector at e defined, as above, from ξ^a and η^a. This $C^m{}_{ab}$ is called the *structure constant tensor* of G. Equations (24) and (25) imply

$$C^m{}_{ab} = C^m{}_{[ab]} \tag{96}$$

$$C^m{}_{n[a}C^n{}_{bc]} = 0 \tag{97}$$

respectively. Eqn. (96) is obvious. For (97), note that $\Big[[\eta, \lambda], \xi\Big]$ is represented by $C^m{}_a\xi^a$, $C^n{}_{bc}\eta^b\lambda^c$.

A (finite-dimensional) *Lie algebra* is a finite-dimensional vector space with a tensor $C^m{}_{ab}$ over that space, satisfying (96) and (97). Thus, every Lie group defines a Lie algebra.

Set $g_{ab} = -C^m{}_{na}, C^n{}_{mb}$, a tensor in G at e. By Theorem 19, we obtain an, obviously symmetric, tensor field g_{ab} on G. This is called the *invariant metric* of the Lie group G. (It many not be a metric as we have defined this term, for g_{ab} may not be invertible.)

Example 56. $SU(2)$ is the Lie group of all complex, unitary, 2×2 matrices with unit determinant. It is underlying manifold is S^3. Hence, the tangent bundle of S^3 is a product. Choose a contravariant vector at e in $SU(2)$, and, by Theorem 19, obtain a nowhere vanishing vector field on S^3. This is called the Hopf fibration of S^3.

Exercise 140. Let G and G' be Lie groups. Show that $G \times G'$, considered as a product of manifolds and a direct sum of groups, is a Lie group.

Exercise 141. Prove that the structure constant tensor of a commutative group vanishes.

Exercise 142. Find all two-dimensional Lie algebras.

Exercise 143. Show that, in the definition of a Lie group, it suffices to require that the mapping from $G \times G$ to G which sends (α, β) to $\alpha\beta^{-1}$ be smooth.

Exercise 144. Find a Lie group whose manifold is the torus; the cylinder.

23. Groups of Motions

We have seen in Sect 20 that a contravariant vector field on a manifold M yields a collection Γ_t (t a number) of diffeomorphisms on M, where these diffeomorphisms satisfy $\Gamma_t \cdot \Gamma_w = \Gamma_{t+w}$ and $\Gamma_0 =$ identity on M. In other words, the additive group of real numbers is realized as a group of diffeomorphisms on M. (More generally, the collection of all diffeomorphisms on a manifold M forms a group, with composition as the group operation. Above, we obtain a subgroup isomorphic with the additive group of reals.) One can consider more general groups realized as diffeomorphisms on a manifold. We briefly describe this type of situation.

Let G be a lie group, and M a manifold. Consider a smooth mapping $\Gamma : \ G \times M \to M$. Writing $\Gamma_\alpha(p)$ for $\Gamma(\alpha, p)$ (so, for each α in G, Γ_α is a smooth mapping from M to M), we require that composition of these mappings reflect the group operation in G, i.e., we require that

$$\Gamma_\alpha \cdot \Gamma_\beta = \Gamma_{\alpha\beta} \qquad \Gamma_e = \text{identity} \tag{98}$$

A *group of motions* on M consists of a Lie group G and a smooth mapping $\Gamma : G \times M \to M$ satisfying (98). By the same argument used in this situation in Sects. 20 and 22, we see that each is a diffeomorphism on M.

Example 57. Every Lie group defines a group of motions on itself.

Example 58. A group of motions on a manifold M defines a group of motions on TM.

Example 59. The Poincaré group is a group of motions on Minkowski space (a 4−manifold).

Example 60. A complete contravariant vector field on M defines, as in Sect 20, a group of motions on M, where G is the additive group of real numbers.

Example 61. The Lie group of nonsingular linear mapping on a finite-dimensional vector space is a group of motions on that vector space (considered as a manifold).

A group of motions is said to be *effective* if $\Gamma_\alpha =$ identity only when $\alpha = e$, and *transitive* if, for any two points p and q of M, there is an α in G such that $\Gamma_\alpha(p) = q$.

Let G be a group of motions on M. Intuitively, a contravariant vector in G at e represents "an element of G differing infinitesimally from e". Hence, it defines an "infinitesimal motion on M". That is, it takes "each point of M to a nearby point". Thus, we should have a contravariant vector at each point of M, i.e., a contravariant vector field on M. Thus, a contravariant vector in G at e should define a contravariant vector field on M. We now define this field.

Let ξ^a be a contravariant vector in G at e. Fix a point p of M. Let $\psi : G \to G \times M$ be defined by $\psi(\alpha) = (\alpha, p)$, Then ψ sends e to the point (e, p) of $G \times M$. Hence, ψ sends ξ^a to some contravariant vector in $G \times M$ at (e, p). But $\Gamma : G \times M \to M$ sends (e, p) to p. Hence, $(\Gamma \vec{\cdot} \psi)_e \xi^a$ is a contravariant vector in M at p. Repeating for each p, we obtain a contravariant vector field on M. Thus, each ξ in G at p defines a field ξ'^a on M.

Let ξ^a and η^a be contravariant vector in G at e. We shall show that

$$[\xi, \eta]' = \mathscr{L}_{\xi'} \eta' \tag{99}$$

That is if the bracket of ξ^a and η^a (in G) is taken to a field on M, the result is the same as the Lie derivative of η'^a in the ξ'^a–direction (i.e., Lie derivatives of fields in M). Take ξ^a as a vector field on G, and let γ_t (t a number) be the corresponding 1–parameter group of motion on G. Then, evidently, $\Gamma_{\gamma_{t(a)}}$ is a 1–parameter group of motions on M. Consider η^a as a contravariant vector field on G. Then, for each t, γ_t takes η^a to some other contravariant vector field on G. Let κ^a be that field evaluated at e. Then κ'^a is the vector field on M obtained by acting on η'^a with the diffeomorphism $\Gamma_{\gamma_{t(t)}}(e)$ on M. Taking d/dt of this equality at $t = 0$, and using (93), Eqn, (99) follows.

To summarize, a group of motions on M (where G is k–dimensional) defines a k–dimensional vector space of contravariant vector fields on M. This vector space is naturally isomorphic to the tangent space to G at e. Lie brackets in G correspond to Lie derivatives of these vector fields) in M. Thus, if a Lie group is realized as a group of motions on M, its Lie algebra is realized as a Lie algebra of contravariant vector fields on M. Note that Sect. 20 is the special case when G is one-dimensional.

Example 62. For the Poincaré group as a group of motions on Minkowski space, the corresponding vector fields are the Killing vectors on Minkowski space.

Example 63. G is effective if and only if the mapping from the tangent space to G at e to vector fields on M is one-to-one. For G to be transitive, it is necessary that $\dim G \geq \dim M$.

Exercise 145. For the translations and rotations on the plane (a group of motions), find the corresponding vector fields and their Lie derivatives explicitly.

24. Dragging Along

Let ξ^a be a contravariant vector field on the manifold M. Then, through each point of M, there passes an integral curve of ξ^a. Now suppose we are given a tensor at a point of M. We wish to transport that tensor along the integral curve of ξ^a (i.e., define a tensor at each point of the curve) by the requirement that the Lie derivative of the tensor in the ξ^a–direction be zero. From the concomitant expression (52), we see that this amounts to solving a certain ordinary differential equation. But we have a method, Theorem 16, for solving ordinary differential equations (or, at least, for asserting existence and uniqueness of solutions). Thus, we wish to arrange matters so that transporting the tensor amounts to finding integral curves of some vector field.

It is convenient to first discuss this problem in a somewhat more general setting. Let B be one of the tensor bundles over M. Let $\gamma : I \to M$ be a curve in M. Then a curve $\tilde{\gamma} : I \to B$ is said to be a *lifting* of γ if $\gamma = \pi \cdot \tilde{\gamma}$. Thus, if $\tilde{\gamma}$ is a lifting of γ, then, for each t in I, $\tilde{\gamma}(t)$ is a point of the fibre over $\gamma(t)$. So a lifting of γ amounts to selecting, for each t, some point in the fibre over $\gamma(t)$. In other words, a lifting of γ is essentially a choice, for each t, of a tensor in M, at $\gamma(t)$. Clearly, a lifting represents a specification of how to transport a tensor along γ.

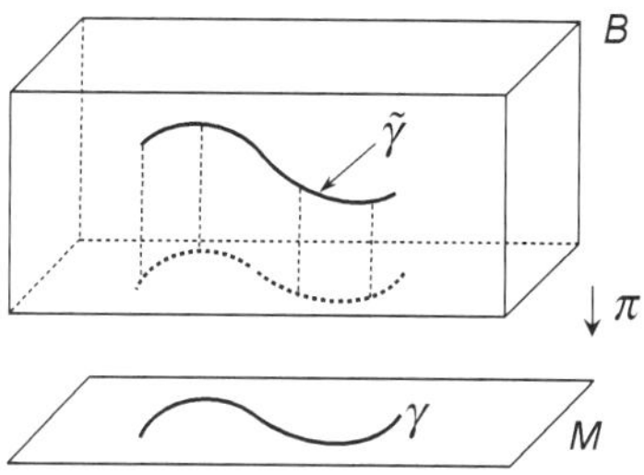

Now let ξ^a be a contravariant vector field in M, and B one of the tensor bundles over M. In order to transport tensors along the integral curves of ξ^a, we must obtain a lifting of each integral curve of ξ^a. A contravariant vector field $\tilde{\xi}^a$ in B will be said to be a *lifting* of ξ^a if, for each point P of B, $\vec{\pi}_P \tilde{\xi}^a = \xi^a$. How unique is the lifting of a vector field? If $\tilde{\xi}^a$ is a lifting of ξ^a, we can add to $\tilde{\xi}^a$ at P any other vector λ^a with $\vec{\pi}_P \lambda^a = 0$. But, as we have seen, there is a k–dimensional vector space of such vectors at P (where k = dimension of the fibres): these are the vertical vectors which lie in the fibres. Geometrically, if ξ^a at p connects p to the "infinitesimally displaced point q", then $\tilde{\xi}^a$ at P (where P is in the fibre over p) must connect P with "any infinitesimally displaced point (of B) in the fibre over q." The choice of the point of the fibre over q is the same as the choice of a λ^a lying within the fibres.

Liftings of vector fields and curves are related by the – intuitively clear – statement:

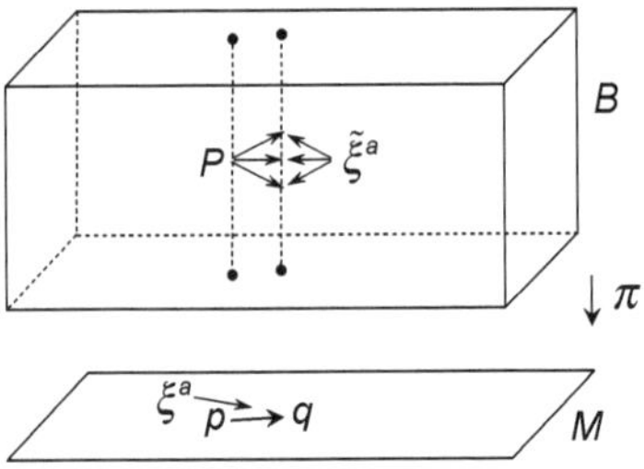

Theorem 20. Let ξ^a be a contravariant vector field on M, B one of the tensor bundles over M, and $\tilde{\xi}^a$ a lifting of ξ^a. Then each integral curve of $\tilde{\xi}^a$ (in B) is a lifting of an integral curve of ξ^a (in M).

Proof: Let $\gamma : \ I \to B$ be an integral curve of $\tilde{\xi}^a$, and let $\pi \cdot \tilde{\gamma} : \ I \to M$ be the corresponding curve in M. We must show that $\pi \cdot \tilde{\gamma}$ is an integral curve of ξ^a. But this is immediate, for the tangent vector to $\pi \cdot \tilde{\gamma}$ at t is $(\overrightarrow{\pi \cdot \tilde{\gamma}})_t = \vec{\pi}_P \cdot \vec{\tilde{\gamma}}_t i^a = \vec{\pi}_P \tilde{\xi}^a = \xi^a$, where $P = \tilde{\gamma}(t), \ p = \pi(P)$.

We now return to the original problem, that of obtaining a mode of transport using the Lie derivative. Let ξ^a be a contravariant vector field in M, B one of the tensor bundles over M. From the discussion above, the problem reduces to that of specifying a particular lifting of ξ^a. For each t, Γ_t is a diffeomorphism on M. Hence, Γ_t defines a diffeomorphism, $\tilde{\Gamma}_t$ on B. Of course, $\tilde{\Gamma}_t$ is fibre preserving: $\pi \cdot \tilde{\Gamma}_t = \Gamma_t \cdot \pi$. Furthermore, $\tilde{\Gamma}_t$ preserves the vector-space structure within each fibre. As in Sect. 23, the one-parameter group of motions $\tilde{\Gamma}_t$ on B defines a contravariant vector field $\tilde{\xi}^a$ in B. By the usual argument, $\tilde{\xi}^a$ is a lifting of ξ^a. Thus, there is a natural lifting of any contravariant vector field in M.

Let $(p, \alpha^{a\ldots c}{}_{b\ldots d})$ be a point of B, so $\alpha^{a\ldots c}{}_{b\ldots d}$ is a tensor in M at p. Then the integral curve of $\tilde{\xi}^a$ through $(p, \ \alpha^{a\ldots c}{}_{b\ldots d})$ is a lifting of the integral curve of ξ^a through p. Thus, the first integral curve defines, at each point of the second integral curve, a tensor in M at that point. In other words, given a tensor in M at p, we obtain a tensor at each point of the integral curve of ξ^a through p. These tensors are said to be the result of *dragging along* of $\alpha^{a\ldots c}{}_{b\ldots d}$.

Clearly, a tensor field on M is equal to the result of dragging along that tensor from any point of M if and only if the Lie derivative of that tensor field in the ξ^a–direction is zero. In this sense, dragging along represents "transport by the requirement that the Lie-derivative in the ξ^a–direction be zero".

Example 64. Geometrically, dragging along of a contravariant vector field η^a can be interpreted as follows. Think of η^a as an infinitesimal displacement. Move each of these two nearby points a parameter-distance t along their respective integral curves of ξ^a. The resulting infinitesimal displacement is the result of dragging along of η^a. Thus, for the ξ^a vanishes at p, so the integral curve through p remains there. But dragging along of a contravariant vector at p results in a rotation of that vector, keeping it at p.

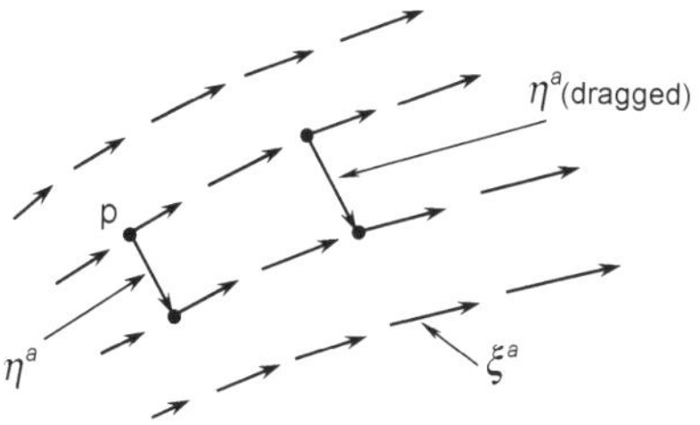

Exercise 146. Prove that the sum of the result of dragging along two tensors is the same as the result of dragging along their sum.

Exercise 147. Let ξ^a Be a contravariant vector field on M, and suppose that Γ_{t_0} = identity, so the integral curves of ξ^a are all closed loops. Prove that the result of dragging along a tensor around one of these closed loops is to leave the tensor invariant.

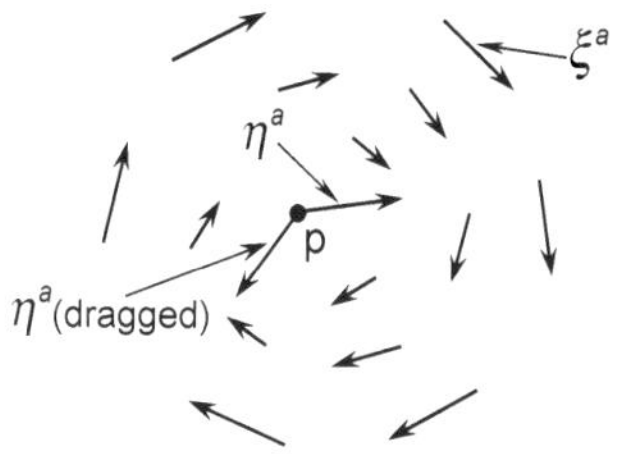

Exercise 148. Find an example of a ξ^a having an integral curve which is a closed loop, such that the result of dragging along a tensor around this closed loop does not leave the tensor invariant.

25. Derivative Operators: Interpretation in the Tensor Bundles

In Sect. 24. we saw that there is a natural lifting of any contravariant vector field $\hat{\xi}^a$ on a manifold to a contravariant vector field $\tilde{\xi}^a$ on each of the tensor bundles of M. Now suppose we have a derivative operator ∇_a on M. What additional structure appears on the tensor bundles? This question is the subject of the present section.

It is convenient to first introduce some notation. Let M be a manifold, B the tensor bundle of tensors with index structure $\alpha^{a...c}{}_{b...d}$. Let $K_a{}^b$ be a tensor field on M, and consider the expression

$$\alpha^{m...c}{}_{b...d}K_m{}^a + \cdots + \alpha^{a...m}{}_{b...c}K_m{}^c - \alpha^{a...c}{}_{m...d}K_b{}^m - \cdots - \alpha^{a...c}{}_{b...m}K_d{}^m \quad (100)$$

We regard (100) as a linear mapping from each fibre to itself. That is, if $(p, \alpha^{a...c}{}_{b...d}$ (α a tensor at p) is a point of B, then expression (100) defines another tensor at p, i.e., another point of B. In other words, if P is a point of B ($p = \pi(P)$), (100) defines an element of the vector space $\pi^{-1}(p)$. That is (100) defines, for each point P of B, a contravariant vector (in B) at P. Thus, we obtain a contravariant vector field on B, which we write $\overline{K}^a$. Thus, we have – a clearly linear – mapping from tensor fields $K_a{}^b$ on M to contravariant vector fields $\overline{K}^a$ on B.

Let ξ^a be a contravariant vector field on M, and let B be one of the tensor bundles over M. Set

$$\hat{\xi}^a = \tilde{\xi}^a + \overline{(\nabla\xi)}^a \quad (101)$$

Since $\vec{\pi}_P\overline{K} = 0$ for any $K_a{}^b$, and since $\tilde{\xi}^a$ is a lifting of ξ^a, it is clear that $\hat{\xi}^a$ is a lifting of ξ^a. Hence, given any tensor $\alpha^{a...c}{}_{b...d}$ at a point p of M, the integral curves of $\hat{\xi}^a$ through $(p, \alpha^{a...c}{}_{b...d})$ defines a tensor at each point of the integral curve of ξ^a trough p. This method of carrying tensors along integral curves is called *parallel transport*. It is clear from (52) that, if ξ^a and $\alpha^{a...c}{}_{b...d}$ are tensor fields on M, then $\alpha^{a...c}{}_{b...d}$ is invariant under parallel transport along the integral curves of ξ^a if and only if

$$\xi^m\nabla_m\alpha^{a...c}{}_{b...d}$$

To summarize, a vector field ξ^a on M can be lifted to $\tilde{\xi}^a$ on B. If, however, there is specified a derivative operator ∇_a on M. then a second lifting $\hat{\xi}^a$, is available.

The lifting $\hat{\xi}^a$ has an important property not shared by $\tilde{\xi}^a$. If ξ^a and η^a are vector fields on M, then, by (29), we have

$$(\widetilde{\xi^a + \mu\eta^a}) = \tilde{\xi}^a + (\underset{\leftarrow}{\pi}\mu)\tilde{\eta}^a - \overline{(\xi D(\mu))}^a \tag{102}$$

Since

$$\begin{aligned}\overline{\nabla(\xi + \mu\eta)}^a &= \overline{\nabla\xi}^a + \overline{(\mu\nabla\eta)}^a + \overline{(\eta D(\mu))}\\ &= \overline{\nabla\xi}^a + (\underset{\leftarrow}{\pi}\mu)\overline{(\nabla\eta)}^a + \overline{(\eta D(\mu))}^a\end{aligned}$$

we have immediately, from (101),

$$(\widehat{\xi^a + \mu\eta^a}) = \hat{\xi}^a + (\underset{\leftarrow}{\pi}\mu)\hat{\eta}^a \tag{103}$$

In other words, the lifting $\hat{\xi}^a$ is linear. Hence, if $p = \pi(p)$, and ξ^a and η^a are contravariant vector fields on M which coincide at p, then $\hat{\xi}^a = \hat{\eta}^a$ at P. That is $\xi^a \rightarrow \hat{\xi}^a$ defines a lifting, not only of contravariant vector fields on M, but also of contravariant vectors at a point of M.

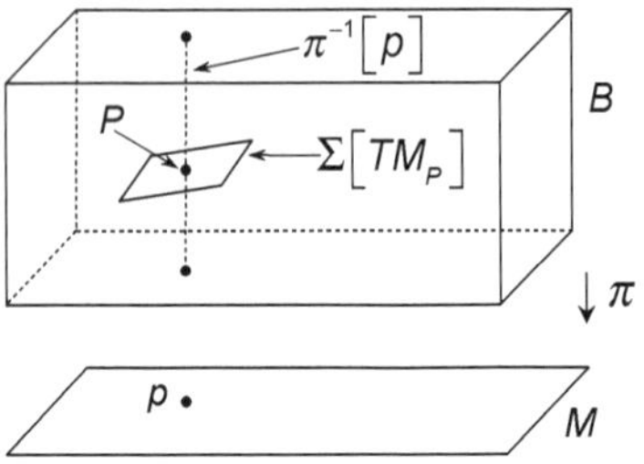

Again, let $p = \pi(P)$. Denote by Σ the mapping from the tangent space to M at p to the tangent space to B at P which sends ξ^a to $\hat{\xi}^a$. Clearly, this mapping is linear, and satisfies $\vec{\pi}_P \cdot \Sigma =$ identity mapping on tangent space to M at p, In particular, Σ is one-to-one. The situation can now be described as follows. Let the fibres of B be k–dimensional. Then $\vec{\pi}_P$ sends a k–dimensional subspace of TB_P (tangent space to B at P) to zero, namely, the subspace of vertical vectors. On the other hand, $\Sigma[TM_P]$ is an n–dimensional subspace of TB_P. Evidently, every element of the $(n + k)$–dimensional vector space TB_P can be written uniquely as the sum of a vertical vector at P and an element of $\Sigma[TM_P]$. Thus, we have decomposed the vector space TB_P into two vector spaces: the vector space of vertical vectors and the vector space $\Sigma[TM_P]$. This $\Sigma[TM_P]$ is called a *horizontal subspace* of TB_P. (In general, a horizontal subspace of TB_P is a subspace such that every element of TB_P can be written uniquely as the sum of an element of the subspace and a vertical vector.)

Thus, a derivative operator on M defines, for each tensor bundle over M, a horizontal subspace of TB_P (P in B). Of course, the derivative operator is uniquely specified by the horizontal subspaces it defies. It is through these horizontal subspaces that a derivative operator on M is represented in the tensor bundles over M.

Exercise 149. Let $\gamma : \ I \rightarrow M$ be a curve, and let ξ^a and ξ'^a be the two vector fields on M both of which have γ as internal curve. Show that parallel transport along γ using ξ^a is identical with parallel transport along γ using ξ'^a. (Hence, parallel transport along a single curve, with no vector field, is well-defined.)

Exercise150. Consider R^2 with metric with components, in the natural chart, $g_{11} = g_{22} = 1$, $g_{12} = g_{21} = 0$. Consider parallel transport with the corresponding derivative operator. Show that the result of parallel transport of any tensor about a closed curve leaves that tensor invariant.

Exercise 151. In what sense does a derivative operator on M define an "inverse" of $\overset{\leadsto}{\pi}$?

Exercise 152. Let ∇_a and ∇'_a be derivative operators on M, and ξ^a a contravariant vector field on M. Let $\hat{\xi}^a$ and $\hat{\xi}'^a$ denote the liftings of ξ^a defined by ∇_a and ∇'_a, respectively. Prove that $\hat{\xi}'^a = \hat{\xi}^a + \overline{K}^a$, where $K_a{}^b = \gamma^b{}_{ma}\xi^m$, and $\gamma^m{}_{ab}$ is the tensor field relating ∇'_a and ∇_a,

Exercise 153. Give an example of two contravariant fields ξ^a and η^a on M which coincide at p, but such that $\overset{\leadsto}{\xi}^a \neq \overset{\leadsto}{\eta}^a$ at P $(p = \pi(P))$.

Exercise 154. Prove that two derivative operators which define the same horizontal subspace are identical.

26. Riemann Tensor: Geometrical Interpretation

Let M be a manifold with derivative operator ∇_a. Then, intuitively speaking "parallel transport of a tensor about a small closed loop changes that tensor by an amount depending on the Riemann tensor". It is in this intuitive sense that the Riemann tensor represents "curvature". Parallel transport about a closed loop (small or otherwise) in the plane leaves the tensor invariant: the plane is flat. Parallel transport of a vector about a closed curve on a S^2 changes that vector: S^2 is curved. (In both cases, "parallel transport" refers to the derivative operator defined by the natural metric.)

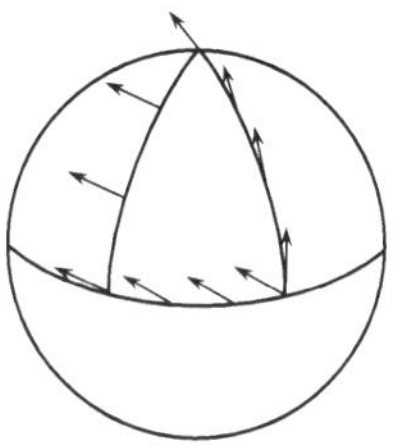

We shell establish the following fact:

$$\mathcal{L}_\xi \hat{\eta}^a - (\widehat{\mathcal{L}_\xi \eta^a}) = -\hat{K}^a \qquad (104)$$

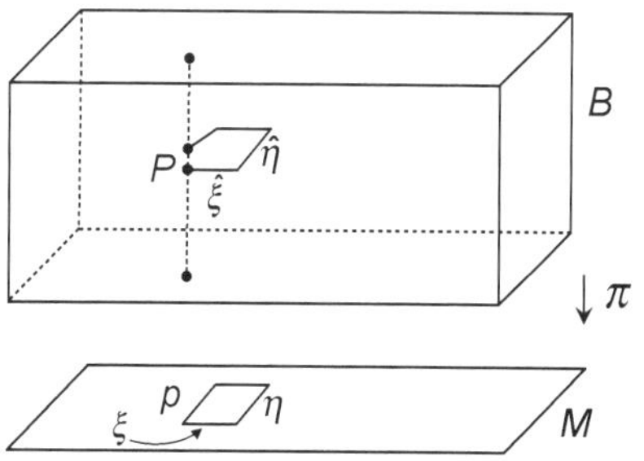

where $K_a{}^b = R_{mna}{}^b \xi^m \eta^n$. Before giving the proof, we interpret (104) geometrically. The commutator in the first term on the left represents passage about an infinitesimal loop in the base space, where the "sides" of this loop are described by the infinitesimal displacements defined by ξ^a and η^a. The second term on the left in (104) represents the corresponding loop in B. The difference between these two quantities represents, therefore, the change in a tensor under parallel transport *ab* out the loop in M. The difference is a vertical vector, i.e., it represents a change in a tensor at p. Thus, Eqn. (104) represents the effect on a tensor under parallel transport about a small closed loop.

The proof of Eqn. (104) consists of the following calculation:

$$[\xi^n\nabla_n(\eta^m\nabla_m\alpha^a{}_b) - \eta^m\nabla_m(\xi^n\nabla_n\alpha^a{}_b)]$$
$$- [\xi^m\nabla_m\eta^n - \eta^m\nabla_m\xi^m]\nabla_n\alpha^a{}_b = 2\xi^m\eta^n\nabla_{[m}\nabla_{n]}\alpha^a{}_b$$
$$= -(\xi^m\eta^n R_{mnp}{}^a)\alpha^p{}_b + (\xi^m\eta^n R_{mnb}{}^p)\alpha^a{}_p$$

27. Geodesics

A certain class of curves, called geodesics, on a manifold with derivative operator are of particular interest.

Let M be a manifold with derivative operator ∇_a. We introduce a contravariant vector field on the tangent bundle of M, TM. If (p, ξ^a) is a point of TM, $\hat{\xi}^a$ is a contravariant vector in TM at the point (p, ξ^a). This vector field is called the *geodesic spray*. Let $\tilde{\gamma} : I \to B$ be an integral curve of the geodesic spray. Then $\gamma = \pi \cdot \tilde{\gamma}$, a curve in M, is called a *geodesic*. Thus, a geodesic is determined by a point of TM, i.e., by a contravariant vector at a point of M. The parameter t along a geodesic is sometimes called an affine parameter. An alternative definition of a geodesic is the following: a geodesic is a curve such that the tangent vector of the curve is parallel transported along the curve.

Exercise 155. Show that the geodesics on R^n with the usual metric are straight lines.

Exercise 156. Show that the geodesics on S^2 with the usual metric are great circles.

Exercise 157. Let ξ^a be a vector field on M. Show that the integral curves of ξ^a are geodesics if and only if $\xi^m \nabla_m \xi^m = 0$.

Exercise158. Let $\gamma : I \to M$ be a geodesic on M (with some derivative operator on M). Let $\mu : I' \to I$ be one function of one variable. Show that $\gamma \cdot \mu : I' \to M$ is a geodesic if and only if there are numbers a and b such that μ sends t to $at + b$.

Exercise 159. Show that, in terms of a chart, a geodesic, $x^1(t), \ldots, x^n(t)$, satisfies $d^2/dt^2 x^i(t) = -\Sigma_{j,k} \Gamma^i_{jk}(x(t)) \frac{dx^j}{dt} \frac{dx^k}{dt}$, where $\Gamma^i_{jk}(x)$ is the connection defined by the derivative operator.

Exercise160. Show that two derivative operators on M which define precisely the same geodesics are identical.

Exercise 161. Let g_{ab} be a metric on M, and let $\gamma : I \to M$ be a geodesic of the corresponding derivative operator. Show that $\xi^a \xi^b$, g_{ab} is constant along the geodesic, where ξ^a is the tangent vector of the curve.

Exercise 162. Consider a group of motions on M such that each Γ_α(α in G) takes the derivative operator to itself. Show that Γ_α takes each geodesic to another geodesic.

28. Submanifolds

Roughly speaking, a (k–dimensional) submanifold of a manifold M Is a (k–dimensional) surface in M.

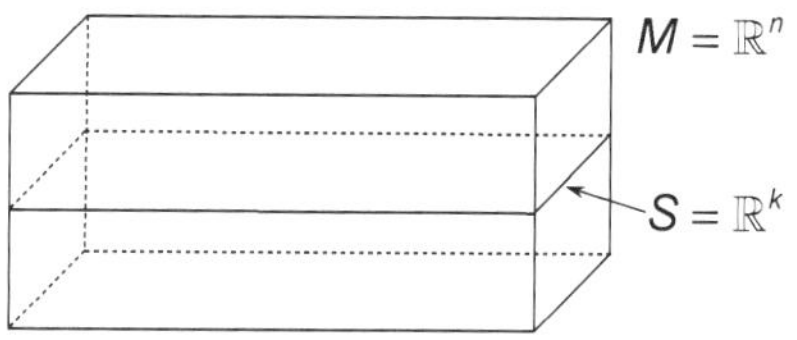

Let $M = R^n$. Denote by S the collection of all points of M for which $x^{k+1} = x^{k+2} = \cdots = x^n = 0$. Clearly, this S is a k–dimensional hyperplane in the n–dimensional Euclidean space M. This is one example of a submanifold. More generally, a submanifold is a subset which, locally, looks like the example above.

Let M be an n–dimensional manifold. A subset S of M is called a k–dimensional *submanifold* if, for each point p of S, there is a chart $(U\varphi)$ of M, with p in U, such that $\varphi[U \cap S]$ consists precisely of the points of $\varphi[U]$ with $x^{k+1} = \cdots = x^n = 0$. In other words, S is a k–dimensional submanifold if, for each point p of S, there is a chart containing p such that S intersects U in precisely the points of U at which the last $(n-k)$ coordinates all vanish. The number $(n-k)$ is called the *co-dimension* of the sub manifold S.

Example 65. Let S be the subset of R^n consisting of points with $(x^1)^2 + \cdots + (x^n)^2 = 1$. Then S is an $(n-1)$–dimensional submanifold of R^n. The co-dimension is one.

Example 66. Let $M = M' \times M''$. Fix a point p'' of M'', and let S consist of all points of M of the form (p', p''), Then S is a submanifold of M, of dimension n' (= dimension of M'), and co-dimension n'' (= dimension of M'').

Example 67. The four figures below represent subset s of R^2. The first three are not submanifolds, for they do not satisfy the required condition at p. The last is a submanifold (dimension 1).

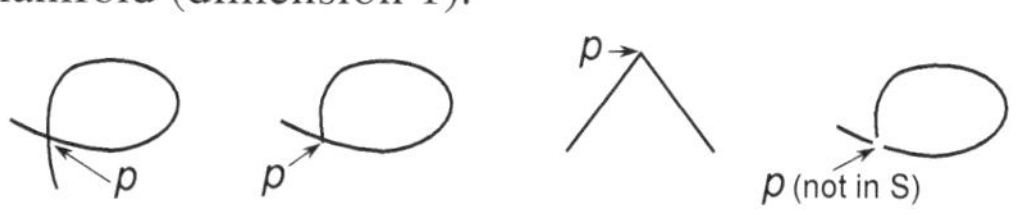

In the last figure, the "intersection point" p is not a point of S, and so no conditions need be satisfied near p. But, for each point of S, the submanifold condition is satisfied.

Example 68. The image of a cross section of a vector bundle is a sub manifold of the bundle.

Example 69. Every open subset of a manifold is an ($n-$dimensional) submanifold. Every point of a manifold is a ($0-$dimensional) submanifold.

Let S be a submanifold of M, so S is a set. We introduce some charts on S. If (U, φ) is a chart on M, with U S consisting of the points of U with all the last $(n - k)$ coordinates zero, then $x^1, \ldots, x^k$ are functions on $U \cap S$. Thus, $U \cap S$ is the subset of S, and $x^1, \ldots, x^k$ are the coordinates on this subset. This is a chart. (Exercise 163. Verify that these charts satisfy $M1 - M4$.) Thus, a $k-$dimensional submanifold S of M itself has the structure of a $k-$dimensional manifold. A submanifold of M represents another manifold which "sits inside" M.

Let S be a sub manifold of M, and let $\psi : S \to M$ be the identity mapping (a point of S, which is also a point of M, is taken by ψ to that point of M). Since S has the structure of a manifold, we can consider ψ as a mapping of manifolds. It is clear from the definitions that ψ is one-to-one, is smooth, and that $\vec{\psi}_p$ (p in S) is one-to-one). (ψ is one-to-one because it is the identity. ψ is smooth because the definition of the manifold structure of S gives precisely the conditions for the characterization of a smooth mapping in terms of charts. $\vec{\psi}_p$ is one-to-one because, if (U, φ) is a chart on M which defines a corresponding chart on S, and if ξ^a is a vector in S at p with components $(\tilde{\xi}^1, \ldots, \tilde{\xi}^k)$, then $\vec{\psi}_p, \xi^a$ is the vector at p in M with components $(\tilde{\xi}^1, \ldots, \tilde{\xi}^k, 0, \ldots, 0)$.)

There are no suppresses here. A submanifold is a smooth surface which does not come back near it itself, and, in particular, does not intersect itself. Such a subset inherits a manifold structure from M, and the mapping from S to M becomes smooth.

Exercise 164. Find an example in which the intersection of two submanifolds is not a submanifold: in which the union of two submanifolds is not a submanifold.

Exercise 165. Let $\psi : M \to M'$ be smooth, and let S be a sub manifold of M. Find an example in which $\psi[S]$ is not a submanifold of M'.

Exercise 166. Let S be a sub manifold of M, T a sub manifold of S (so T, as a subset of S, is also a subset of M). Show that T is a submanifold of M.

Exercise 167. Let $\gamma : I \to M$ be a curve with nowhere vanishing tangent vector. Show that, for each t_0 in I, there is an interval I' containing t_0 such that $\gamma[I']$ is a sub manifold of M. Find a counterexample to show this is false without the condition "with nowhere vanishing tangent vector". (Hint: Use third figure of Example 67.)

Example 70. Each fibre is a sub manifold of a vector bundle.

29. Tangents and Normals to Submanifolds

Let $\psi : S \to M$ be a k–dimensional submanifold of an n–dimensional manifold M. In this section, we relate the tensors on S to the tensors on M. [For the purposes of this section, it is convenient to regard S as a manifold separate from M, and not as a subset of M.]

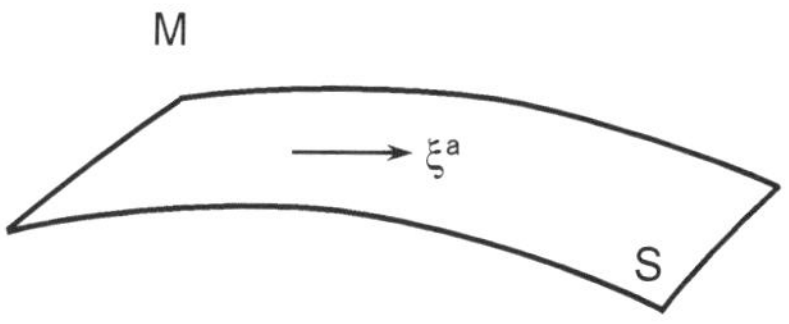

Fix, once and for all, a point p' of S, and set $p = \psi(p')$, a point of M. If ξ'^a is any contravariant vector in S at p', then $\xi^a = \vec{\psi}_{p'}\xi'^a$ is a contravariant vector in M at p. A vector ξ^a such that $\xi^a = \vec{\psi}_{p'}\xi'^a$ for some ξ'^a will be said to be *tangent* to S. [Regarding S as a subset of M, this is the usual, intuitive, geometrical notion of tangency.] Evidently, the contravariant vectors at p tangent to S form a k–dimensional subspace of the n–dimensional tangent space to M at p. A contravariant vector μ_a in M at p will be said to be *normal* to S if $\xi^a\mu_a = 0$ for all ξ^a tangent to S. The normal vectors to S form an $(n-k)$–dimensional subspace of the cotangent space to M at p. Clearly, ξ^a in M at p is tangent to S if and only if $\xi^a\mu_a = 0$ for every μ_a normal to S. Thus, a k–dimensional submanifold S of M defines, at each point of $\psi[S]$, a k–dimensional subspace of the tangent space to M and $(n-k)$–dimensional subspace of the cotangent space to M.

More generally, let $\alpha^{a\ldots c}{}_{b\ldots d}$ be any tensor in M at p. A contravariant index of α, e.g., "c", will be said to be *tangent* to S if $\mu_c\alpha^{a\ldots c}{}_{b\ldots d} = 0$ for all μ_c at p normal to S. Similarly, a covariant index of α, e.g. "b", will be said to be *normal* to S if $\xi^b\alpha^{a\ldots c}{}_{b\ldots d} = 0$ for all ξ^a at p tangent to S.

Exercise 168. Verify that, for "a" in $\alpha^{a\ldots c}{}_{b\ldots d}$ tangent to S, and "a" in $\beta^{p\ldots q}{}_{m\ldots a}$ normal to S, $\alpha^{a\ldots c}{}_{b\ldots d}\beta^{p\ldots q}{}_{m\ldots a} = 0$.

Exercise 169. Prove that μ_a in M at p is normal to S if and only if $\underleftarrow{\psi_p}\,\mu_a = 0$.

Exercise 170. Let f be a scalar field on M such that f is constant on $\psi[S]$. Prove that, at each point of $\psi[S]$, $D(f)_a$ is normal to S.

Exercise 171. Let $\gamma : I \to M$ be a curve in M which remains in $\psi[S]$. Prove that, for each t, the tangent vector to γ is tangent to S.

30. Metric Submanifolds

Let $\psi : S \to M$ be a submanifold of M. [We continue to regard a submanifold S as separate from M, and not a subset.] Suppose we have a metric g_{ab} on M. Then, since we can raise and lower indices of tensors on M, additional structure is induced on M and S.

Set $h'_{ab} = \underleftarrow{\psi} g_{ab}$, a symmetric tensor field on S. If this h'_{ab} on S has an inverse, i.e., if h'_{ab} is a metric on S, then S will be said to be a *metric* sub manifold.

Example 71. Let $S = R^1$ (coordinate x), $M = R^2$, and let $\psi : S \to M$ be $\psi(x) = (x, x)$. Let g_{ab} be the metric on M whose components, in this coordinate system, are $g_{11} = 1$, $g_{22} = -1$, $g_{12} = g_{21} = 0$. Then $\underleftarrow{\psi} g_{ab} = 0$, so S is not a metric submanifold.

We now prove the following (not very important, but illustrative) result: S is a metric sub manifold if and only if, for each point p of $\psi[S]$, the only contravariant vector ξ^a at p with ξ^a tangent to S and $\xi^a = g_{ab}\xi^b$ normal to S is $\xi^a = 0$. To prove this, note that $\underleftarrow{\psi}_{p'}\, g_{ab}$ is not invertible if and only if it has a nullspace, i.e., if and only if $\eta'^a \underleftarrow{\psi}_{p'}\, g_{ab} = 0$ for some nonzero vector η'^a in S at p' (where $p = \psi(p')$). But, if η'^a satisfies these conditions, then $\xi^a = \vec{\psi}_{p'}\eta'^a$ is tangent to S, while $\underleftarrow{\psi}_{p'}\, g_{ab}\xi^b = \underleftarrow{\psi}_{p'}\, (g_{ab}\, \vec{\psi}_{p'}\, \eta'^a) = \eta'^b\, \underleftarrow{\psi}_{p'}\, g_{ab} = 0$, so ξ_a is normal to S. Conversely, if ξ^a is tangent to S, with ξ_a normal to S, then the η'^a in S at p' such that $\vec{\psi}_{p'}\, \eta'^a = \xi^a$ is in the nullspace of $\underleftarrow{\psi}_{p'}\, g_{ab}$, so S is not a metric submanifold.

Suppose for a moment that g_{ab} were positive-definite. Then, if S were not a metric submanifold, there would exist a nonzero ξ^a tangent to S with Σ_a normal to S. But this would imply $\xi^a\xi_a = 0$, which is impossible for a nonzero vector and positive-definite metric. We conclude that every submanifold of a manifold with positive-definite metric is a metric submanifold.

Let g_{ab} be a metric on M, $\psi : S \to M$ a metric submanifold of M. Let $\alpha^{a\ldots c}{}_{b\ldots d}$ be a tensor in M at a point p of $\psi[S]$. We say that an index of α is *normal* to S if, when that index is lowered (if necessary; with the metric) it is normal to S (as in Sect. 29). Similarly, an index of α is *tangent* to S if, when that index is raised, it is tangent to S. An index of a tensor cannot be simultaneously normal and tangent unless the tensor itself vanishes.

Let S be a k–dimensional metric submanifold of M. What do we get at a point p of $\psi[S]$? We have the k–dimensional subspace of the tangent space at p consisting of vectors tangent to S. We also have the $(n-k)$–dimensional subspace of the cotangent space of p consisting of vectors normal to S. But, since M has a metric, we can raise the indices of the vectors in the latter subspace. Thus, we obtain an $(n-k)$–dimensional subspace of the tangent space at p consisting of vectors normal to S. These two subspace of the tangent space at p intersect only at the origin. Furthermore, their dimensions – k and $(n-k)$ – add up to n, the dimension of the tangent space at p. Thus, we have decomposed the tangent space at p into the direct sum of two subspaces. It should be possible to introduce projections into the two subspaces, etc. We now do this.

As usual, fix p' in S, and set $p = \psi(p')$. Since $\underset{\leftarrow}{\psi}\, g_{ab}$ is a metric on S, the inverse metric, h'^{ab}, exists (on S). Set $h^{ab} = \vec{\psi}_{p'}\, h'^{ab}$, a symmetric tensor in M at p. It is immediate from the definition that both indices of h^{ab} are tangent to M. We raise and lower indices of h^{ab} using the metric g_{ab} of M. We now prove the following: a contravariant vector ξ^b at p is tangent to S if and only if $\xi^a = h^a{}_b\,\xi^b$. The "if" part is clear. To prove the converse, let ξ^a be tangent to S, so $\xi^a = \vec{\psi}_{p'}\,\xi'^a$. then

$$\underset{\leftarrow}{\psi}_{p'}\xi_a = \underset{\leftarrow}{\psi}_{p'}\,(g_{ab}\xi^b) = \underset{\leftarrow p'}{\psi}\,(g_{ab}\vec{\psi}\xi'^b) = \xi'^b \underset{\leftarrow}{\psi}\, g_{ab}$$
$$= h'_{ab}\xi'^b$$

Hence,

$$\xi^a = \vec{\psi}_{p'}\xi'^a = \vec{\psi}_{p'}\,(h'^{ab}\, h'_{bc}\xi'^c) = \vec{\psi}_{p'}\,(h'^{ab}\, \underset{\leftarrow}{\psi}_{p'}\xi_b)$$
$$= (\vec{\psi}_{p'}\, h'^{ab})\,\xi_b = h^a{}_b\xi^b$$

Similarly, a covariant μ_a in M at p is normal to S if and only if $\mu_a\, h^a{}_b = 0$. [Proof: The "only if" is clear. Let $\mu_a\, h^a{}_b = 0$ Then, for every ξ^a tangent to S, $\mu_a\,\xi^a = \mu_a\,(h^a{}_b\,\xi^b) = 0$. So, μ_a is normal to S.] By the same argument, we see that an index, e.g., "a". of a tensor $\alpha^{\cdots a \cdots}$ at p is tangent to S if and only if

$$h^a{}_b\alpha^{\dots b\dots} = \alpha^{\dots a\dots} \tag{105}$$

and normal to S if and only if

$$h^a{}_b\alpha^{\dots b\dots} = 0 \tag{106}$$

In particular, $h^a{}_b h^b{}_c = h^a{}_c$. Hence, setting $k_{ab} = g_{ab} - h_{ab}$, we have

$$h^a{}_b k^b{}_c = h^a{}_b(\delta^b{}_c - h^b{}_c) = h^a{}_c - h^a{}_c = 0 \tag{107}$$

and

$$k^a{}_b k^b{}_c = (\delta^a{}_b - h^a{}_b)(\delta^b{}_c - h^b{}_c) = \delta^a{}_c - 2h^a{}_c + h^a{}_c \qquad (108)$$
$$= \delta^a{}_c - h^a{}_c = k^a{}_c$$

Evidently, ξ^a satisfies $\xi^b k^a{}_b = 0$ if and only if ξ^a satisfies $\xi^a = \xi^b h^a{}_b$. Thus, reversing "tangent" and "normal", statements about h_{ab} become statement about k_{ab}. We summarize:

Theorem 21. Let g_{ab} be a metric on M, $\psi : S \to M$ a metric submanifold of M, and $p = \psi(p')$ a point of M. Then there are tensors h_{ab} and k_{ab} in M at p with the properties:

1. $h_{ab} + k_{ab} = g_{ab}$
2. Both indices of h_{ab} are tangent to S; both indices of k_{ab} are normal to S.
3. An index "a" of a tensor $\alpha^{\cdots a \cdots}$ at p is tangent to S if and only if $h^a{}_b \alpha^{\cdots b \cdots} = \alpha^{\cdots a \cdots}$, which, in turns, holds if and only if $k^a{}_b \alpha^{\cdots b \cdots} = 0$.
4. An index "a" of a tensor $\alpha^{\cdots a \cdots}$ at p is normal to S if and only if $k^a{}_b \alpha^{\cdots b \cdots} = \alpha^{\cdots a \cdots}$, which in turn, holds if and only if $h^a{}_b \alpha^{\cdots b \cdots} = 0$.
5. $h_a{}^b h_b{}^c = h_a{}^c$, $k_a{}^b k_b{}^c = k_a{}^c$ (projections operators).

A given index of a tensor at p may, of course, be neither tangent nor normal to S. But Theorem 21 has the useful consequence that any tensor may be decomposed into tensors all of whose indices are either tangent or normal. We might as well take all indices contravariant. We do the case with two indices as an example: the n–index case should be clear:

$$\alpha^{ab} = \alpha^{mn}\delta^a{}_m\delta^b{}_n = \alpha^{mn}(h^a{}_m + k^a{}_m)(h^b{}_n + k^b{}_n) \qquad (109)$$
$$= \alpha^{mn}h^a{}_m h^b{}_n + \alpha^{mn}k^a{}_m h^b{}_n = \alpha^{mn}h^a{}_m k^b{}_n$$
$$+ \alpha^{mn}k^a{}_m k^b{}_n$$

To summarize, the tensors at a point of M on a metric submanifold have the nicest decomposition one could have possible asked for.

Exercise 172. what goes wrong with above if S is not a metric submanifold?

Exercise . 173 Prove that $h^a{}_a = k$ (=dimension of S); that $k^a{}_a = (n - k)$

Exercise 174. Consider Example 65. Let $M = R^n$ have metric with components $g_{11} = \ldots = g_{nn} = 1$, all others zero. Show that S is a metric sub manifold, and find h_{ab} and k_{ab}.

Exercise 175. Show that, in co-dimension one, $k_{ab} = \pm\xi_a\xi_b$ where ξ_a is a normal vector to S. When will the plus sign be applicable?

Exercise 176. Let the submanifold S have dimension one, so S is the image of a curve. Show that $h_{ab} = \alpha\eta_a\eta_b$, where α is a scalar and η_a is the tangent to the curve.

Exercise 177. Find an example of a metric submanifold S such that the metric h'_{ab} on S is indefinite.

Exercise 178. Find h_{ab} and k_{ab} for a manifold of co-dimension zero.
Exercise 179. Define the signature of h_{ab} and k_{ab}, and prove a theorem relating these signatures to that of g_{ab}.

31. Tensor Fields and Derivatives on Submanifolds

Let M be a manifold, and S a subset of M such that S is a $k-$dimensional sub manifold. (Note that we are now taking a submanifold as a subset. We shall allow ourselves to switch from one interpretation to the other (equivalent) one.) A *tensor field on* the submanifold S consists of an assignment, to each point p of S, a tensor (in M) at p, where all the tensor at various points have the same index structure. Thus, if B is the appropriate tensor bundle over M, a tensor field on S is a mapping $\Lambda : S \to B$ such that $\pi \cdot \Lambda =$ identity on S. (Example: When $S = M$, tensor fields are cross sections.) A tensor field on S is said to be *smooth* if the corresponding mapping of manifolds, Λ, is smooth.

Smoothness can be expressed in terms of components. Let (U, φ) be a chart on M such that at $U \cap S$ consisting precisely of the points of U with coordinates $x^{k+1} = \ldots = x^n = 0$. Let $\alpha^{a \ldots c}{}_{b \ldots d}$ be a tensor field on the submanifold S. Then the components of $\alpha^{a \ldots c}{}_{b \ldots d}$ (with respect to this chart) are functions of the k variables $x^1, \ldots, x^k$ (for $\alpha^{a \ldots c}{}_{b \ldots d}$ is only defined on S, were the last $(n-k)$ coordinates vanish). Evidently, $\alpha^{a \ldots c}{}_{b \ldots d}$ is smooth if and only if these functions of k variables are C^∞ for every such chart. (Exercise 180. Prove this.)

Clearly, addition, outer products, contraction, and index substitution on smooth tensor fields on the submanifold S result in smooth tensor fields on the submanifold S. As usual, we normally omit the adjective "smooth". (Differential geometry is the study of smoothness.)

There is a chance of confusion in the terminology here. If we consider S as a manifold in its own right, then we have tensor fields on this manifold. These are different from tensor fields on the submanifold S, as defined above. We keep this distinction by calling the former "tensor fields on the manifold S," and the latter "tensor fields on the submanifold S." (Exercise 181. Find a natural, one-to-one correspondence between contravariant tensor fields on the manifold S and contravariant tensor fields on the submanifold S all of whose indices are tangent to S.)

The remarks above apply, of course, to submanifolds in general, i.e., without further structure. The next logical step would be to introduce the notion of a derivative operator on the submanifold S, and then prove that derivative operator on M induces a natural derivative operator on the submanifold S. This is straightforward, devoid of surprises, and ... not particularly useful. We shall, therefore, omit the discussion of these intermediate structures, and proceed directly to the most common and important case: metric submanifolds. (Exercise 182. Define a derivative operator on the submanifolds S. Show that a derivative operator on M indices a natural derivative operator on each submanifold S.)

Let M be a manifold, g_{ab} a metric on M, and S, a subset of M, a k–dimensional metric sub manifold of M. Then g_{ab}, h_{ab}, and k_{ab} (Sect. 30) are tensor fields on the submanifold S. (Exercise 183. Prove smooth.) Let ∇_a be the derivative operator (on M) associated with g_{ab}. Finally, let $\alpha^{b...c}{}_{r...s}$ be any (smooth) tensor field on the submanifold S. It is clear from the characterization in terms of charts on the previous page that there exists a tensor field $\tilde{\alpha}^{b...c}{}_{r...s}$ on M such that, at points of S, $\tilde{\alpha}^{b...c}{}_{r...s}$ coincides with $\alpha^{b...c}{}_{r...s}$. (In fact, it is only immediately clear that the is true locally. That's all we need. It is true as stated, i.e., globally.) We shall call $\tilde{\alpha}^{b...c}{}_{r...s}$ an extension of $\alpha^{b...c}{}_{r...s}$ to M. Evidently, if $\tilde{\alpha}^{b...c}{}_{r...s}$ and $\tilde{\alpha}'^{b...c}{}_{r...s}$ are two extensions of $\alpha^{b...c}{}_{r...s}$ then $\tilde{\alpha}^{c...c}{}_{r...s} - \tilde{\alpha}'^{b...c}{}_{r...s}$ vanishes on S, so

$$\tilde{\alpha}^{b...c}{}_{r...s} - \tilde{\alpha}'^{b...c}{}_{r...s} = \mu\beta^{b...c}{}_{r...s} + \ldots + \nu\gamma^{b...c}{}_{r...s} \tag{110}$$

where all fields are on M, and the scalar fields vanish on S. (Exercise 184. By using components, establish the existence of such an expansion (110).) Now, let p be a point of S, ξ^a a vector at p tangent to S. We define $\xi^a D_a \alpha^{b...c}{}_{r...s}$ by the equation

$$\xi^a D_a \alpha^{b...c}{}_{r...s} = \xi^a \nabla_a \tilde{\alpha}^{b...c}{}_{r...s} \tag{111}$$

where $\tilde{\alpha}^{b...c}{}_{r...s}$ is an extension of $\alpha^{b...c}{}_{r...s}$. The right side of (111) is independent of the choice of extension, for, using (110),

$$\begin{aligned}\xi^a\nabla_a\tilde{\alpha}^{b...c}{}_{r...s} &= \xi^a\nabla_a\tilde{\alpha}'^{b...c}{}_{r...s} + \mu\xi^a\nabla_a\beta^{b...c}{}_{r...s} + \beta^{b...c}{}_{r...s}\xi^a\nabla_a\mu \\ &+ \cdots + \nu\xi^a\nabla_a\gamma^{b...c}{}_{r...s} + \gamma^{b...c}{}_{r...s}\xi^a\nabla_a\nu = \xi^a\nabla_a\tilde{\alpha}'^{b...c}{}_{r...s}\end{aligned}$$

Thus, $xi^a D_a \alpha^{b...c}{}_{r...s}$ has a "natural" definition only for ξ^a tangent to S. (Intuitively, for tensor fields only defined on S, the derivative can only be taken, meaningfully, along directions along S.) For ξ^a normal to S, set $\xi^a D_a \alpha^{b...c}{}_{r...s} = 0$. Thus (since every vector is a unique linear combination of a vector tangent to S and a vector normal to S) $\xi^a D_a \alpha^{b...c}{}_{r...s}$ is defined for all ξ^a, and is linear in ξ^a. Thus, we obtain a tensor field $D_a \alpha^{b...c}{}_{r...s}$ on the sub manifold S. The index "a" in $D_a \alpha^{b...c}{}_{r...s}$ is, clearly, tangent to S. (This says the same thing: only derivatives in directions tangent to S are meaningful. By fiat, we set derivatives in normal directions equal to zero.) The various properties of ∇_a carry over immediately (e.g., from (111)) to D_a. Thus, we have the Leibnitz rule,

$$D_a\left(\alpha^{\cdots}{}_{\ldots}\beta^{\cdots}{}_{\ldots}\right) = \alpha^{\cdots}{}_{\ldots}\, D_a\beta^{\cdots}{}_{\ldots} + \beta^{\cdots}{}_{\ldots}\, D_a\alpha^{\cdots}{}_{\ldots} \tag{112}$$

additivity,

$$D_a\left(\mu^{\cdots}{}_{\ldots} + \nu^{\cdots}{}_{\ldots}\right) = D_a\mu^{\cdots}{}_{\ldots} + D_a\nu^{\cdots}{}_{\ldots} \tag{113}$$

etc.

To summarize, if S is a metric submanifold of M, then with any tensor field, $\alpha^{b\ldots c}{}_{r\ldots s}$, on the sub manifold S, there is associated another tensor field, $D_a\alpha^{b\ldots c}{}_{r\ldots s}$, on the submanifold S, where the index "a" is tangent to S. This D_a is additive and satisfies the Leibnitz rule, and the application of D_a commutes with contraction and index substitution. [Note: $D_a = h_a{}^m\nabla_a$].

We establish one further important property of this D_a. Let $\psi : S \to M$ be a metric submanifold of M. Then, if $\alpha_{b\ldots c}$ is a tensor field on the submanifold S all of whose indices are tangent to S, then $\alpha'_{b\ldots c} = \overleftarrow{\psi}\,\alpha_{b\ldots c}$ is a tensor field on the manifold S. This is, evidently, a one-to-one correspondence between tensor fields on the manifold S and the tensor fields on the submanifold S all of whose indices are tangent to S. Now,

$$D'_a\alpha'_{b\ldots c} = \overleftarrow{\psi}\left(D_a\,\alpha_{b\ldots c}\right)$$

defines an operator D'_a on covariant tensor fields on the manifold S. It is clear from the summary above that this D'_a is, in fact, a derivative operator on the manifold S. We prove that this D'_a is precisely the derivative operator on the manifold S defined by the metric h'_{ab} on the manifold S. The proof is easy:

$$D'_a h'_{bc} = D'_a(\overleftarrow{\psi}\, g_{bc}) = \overleftarrow{\psi}\,(D_a\, g_{bc}) = \overleftarrow{\psi}(h_a{}^m\nabla_m g_{bc}) = \overleftarrow{\psi}(0) = 0$$

Thus, instead of talking of tensor fields on the manifold S, we can talk of tensor fields on the submanifold S all of whose indices are tangent to S. If $\alpha_{b\ldots c}$ is such a tensor field on the submanifold S, then

$$h_b{}^{m}\cdots h_c{}^{n}\, D_a\,\alpha_{m\ldots n} \tag{114}$$

is another. Thus, we have a derivative operator on the manifold S, and this turns out to be precisely the derivative operator defined by the metric on the manifold S. Quite generally, we can ignore the manifold S: tensors and tensor operations on the manifold S can be replaced by certain tensors (those whose indices are tangent to S) and tensor operations on the sub manifold S.

Example 72. Let the curve $\gamma : I \to M$ be a metric submanifold of M. Then a tensor field $\alpha^{b\dots c}{}_{r\dots s}$ on the submanifold $\gamma[I]$ satisfies $D_a \alpha^{b\dots c}{}_{r\dots s} = 0$ if and only if $\alpha^{b\dots c}{}_{r\dots s}$ is parallel transported along the curve γ.

Exercise 185. Considering $\delta^b{}_c$ as a tensor field on the submanifold S, show that $D_a \delta^b{}_c = 0$.

Exercise 186. Prove that $D_a g_{bc} = 0$.

Exercise 187. Prove that $h_b{}^m h_c{}^n D_a h_{mn} = 0$.

Exercise 188. Prove that D_a commutes with raising and lowering indices of tensor fields on the sub manifold S (using g_{ab}).

Exercise 189. Let ξ^a and η^a be vector fields on the submanifold S, with indices tangent to S. Define $\mathcal{L}_\xi \eta = \xi^m D_m \eta^a - \eta^m D_m \xi^a$. (Why is a definition needed?) Show that this expression satisfies all the properties of Lie derivatives of vectors.

Example 73. If $S = M$, all the constructions above reduce to standard differential geometry on M.

Example 74. Let S be a submanifold of M. Then the tensors (in M) at points of S form vector bundles over S (as base space). Note that the dimensions of the fibres depend only on the dimension of M (and, of course, on the rank of the tensors under consideration), are not on the dimension of S. Now suppose S is a metric submanifold of M. Then, D_a induces a horizontal subspace of these bundles. Furthermore, each of these bundles over S is a submanifold of the corresponding bundle over M. The horizontal subspace of the two bundles are related in the obvious way.

32. Extrinsic Curvature

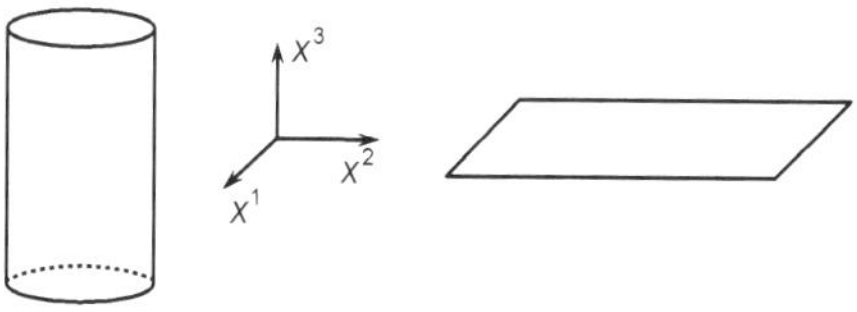

Let $M = R^3$ with the usual positive-definite metric g_{ab}. Consider the two submanifolds of M consisting, on the one hand, of the cylinder $((x^1)^2 + (x^2)^2 = 1)$, and, on the other hand, of the plane $(x^3 = 0)$. Since the metric is positive-definite, these are metric submanifolds. The induced metric h'_{ab} on each submanifold is flat (has vanishing Riemann tensor). (Intuitively, a piece of paper in R^3 can form either a plane or a cylinder.) But, clearly, the cylinder is "curved" in a way that the plane is not. This "curvature" is not an intrinsic property of the geometry of the submanifold, but, rather, expresses the "way in which the submanifold is embedded in M". The embedding of a metric submanifold S in M is described by a certain tensor field on the submanifold S called the extrinsic curvature. It is the extrinsic curvature, and not the intrinsic (local) geometry which distinguishes the cylinder in R^3 from the plane in R^3.

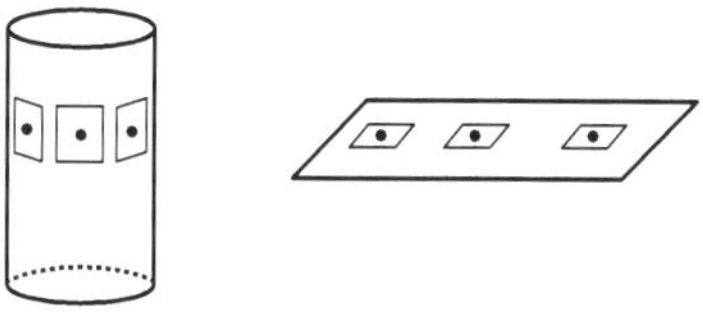

Let S be a metric sub manifold of M. The discussion of Sect. 31 suggests that the quantities $D_a h_{bc}$ and $D_a k_{bc}$ (of course, $D_a h_{bc} = D_a k_{bc} = 0$, so only one needs be considered) should be investigated. Intuitively, $D_a h_{bc}$ represents the "rate of change of the tangent plane to S along S". We might expect this quantity to be zero for the plane (where "all the tangent planes to S at points of S are parallel"), and nonzero for the cylinder (where "the tangent plane to S changes orientation as we move about S").

Thus, $D_a h_{bc}$ does, intuitively, represents the "bending of S in M" – the quantity which distinguishes the plane from the cylinder. We proceed to the study of $D_a h_{bc}$.

The first step in studying any tensor field on the sub manifold S is to project the indices tangent and normal to S (See (109).). this simplification is always possible – and always fruitful. First note that

$$h_b{}^m h_c{}^n D_a h_{mn} = h_b{}^m D_a (h_c{}^n h_{mn}) - h_b{}^m h_{mn} D_a h_c{}^n$$
$$= h_b{}^m D_a h_{cm} - h_b{}^n D_a h_{cn} = 0$$

Furthermore,

$$k_b{}^m k_c{}^n D_a h_{mn} = k_b{}^m D_a (k_c{}^n h_{mn}) - k_b{}^m h_{mn} D_a k_c{}^n$$
$$= k_b{}^m D_a(0) - 0\, D_a k_c{}^n = 0$$

hence, only the "mixed piece" remains:

$$h_b{}^m k_c{}^n D_a h_{mn} = \pi_{abc} \tag{115}$$

This tensor field π_{abc} on the (metric) submanifold S is called the *extrinsic curvature* of S. Note that the first two indices of π_{abc} are tangent to S, the last index normal to S. (One can convince himself intuitively that this is the correct index structure to describe the "bending" of S in M.)

From the results above,

$$D_a h_{bc} = \delta_b{}^m \delta_c{}^n D_a h_{mn} = (h_b{}^m + k_b{}^m)(h_c{}^n + k_c{}^n) D_a h_{mn}$$
$$= (k_b{}^m h_c{}^n + h_b{}^m k_c{}^n)\, D_a h_{mn} = 2\pi_{a(bc)}$$

Hençe, $D_a k_{bc} = -2\pi_a(bc)$. So, the extrinsic curvature completely determines $D_a k_{bc}$ and $D_a k_{bc}$

There is one additional symmetry of π_{abc}, which we now establish. Let ξ^a be a vector field on the submanifold S, normal to S. Then

$$\pi_{[ab]c}\xi^c = \xi^c k_c{}^n h_{[b}{}^m D_{a]} h_{mn} = \xi^n h_{[b}{}^m D_{a]} h_{mn} = h_{[b}{}^m D_{a]} (\xi^n h_{mn}) \tag{116}$$
$$- h_{mn} h_{[b}{}^m D_{a]} \xi^n = -h_{[b}{}^n D_{a]} \xi^n = -h_a{}^m h_b{}^n \nabla_{[m} \tilde{\xi}_{n]}$$

Let $\tilde{\xi}_n$ be an extension of ξ^n. We now "pull back" the right side of (116) using $\underset{\leftarrow}{\psi}$ (where $\psi : S \to M$):

$$\underset{\leftarrow}{\psi}(h_a{}^m h_b{}^n \nabla_{[n} \tilde{\xi}_{n]}) = \underset{\leftarrow}{\psi} \nabla_{[m} \tilde{\xi}_{n]} = \underset{\leftarrow}{\xi} D(\tilde{\xi})_{mn}$$
$$= D(\underset{\leftarrow}{\psi} \tilde{\xi})_{mn} = 0$$

where, for the last equality, we have used the fact that ξ_a is normal to S. Thus, $\pi_{[ab]c}\xi^c$ has all indices tangent to S, and $\underset{\leftarrow}{\psi}(\pi_{[ab]c}\xi^c) = 0$: so $\pi_{[ab]c}\xi^c = 0$. But ξ^c is arbitrary (normal to S, but it makes no difference, since "c" of π_{abc} is normal to S), whence $\pi_{[ab]c} = 0$ that is to say, π_{abc} is symmetric in its first two indices. (Note, incidentally, that a nonzero tensor will never be either symmetric or antisymmetric under interchange of a pair of indices one of which is normal, the other tangent, to S.) Thus, π_{abc} has $(n-k)k(k+1)/2$ independent components.

There is a pretty little formula which gives a good geometrical picture of what the extrinsic curvature is. Consider a geodesic within the sub manifold S, so its tangent vector (which is tangent also to S) satisfies

$$h^a{}_c\eta^b D_b\eta^c = 0 \tag{117}$$

We ask for the curvature of this curve, considered as a curve in M. We have

$$\begin{aligned}\eta^b\nabla_b\eta^c &= \eta^b D_b\eta^c = k^a{}_c\eta^b D_b\eta^c = \eta^b D_b(k^a{}_c\eta^c) \\ &- \eta^b\eta^c D_b k^a{}_c = 0 - \eta^b\eta^c D_b k^a{}_c \\ &= 2\pi_{bc}{}^a\eta^b\eta^c\end{aligned} \tag{118}$$

Thus, although our curve in S is a geodesic in S, it in general has curvature in M. This curvature vector is normal to M, and given by the extrinsic curvature. Evidently, these "curvatures of geodesics" determine π_{abc} completely. Consider again the plane and cylinder. The curves γ_1 and γ_2 are geodesics in their submanifolds, and also geodesics (straight lines) in R^3. Thus, the extrinsic curvature of the plane is zero, while $\pi_{ab}{}^c\eta^a\eta^b$ for η^a tangent to γ_1 in the cylinder is zero. On the other hand, γ_3 is a geodesic in the cylinder, but not a geodesic in R^3. Hence, $\pi_{ab}{}^c\eta^a\eta^b$ for η^a tangent to γ_3 is not zero.

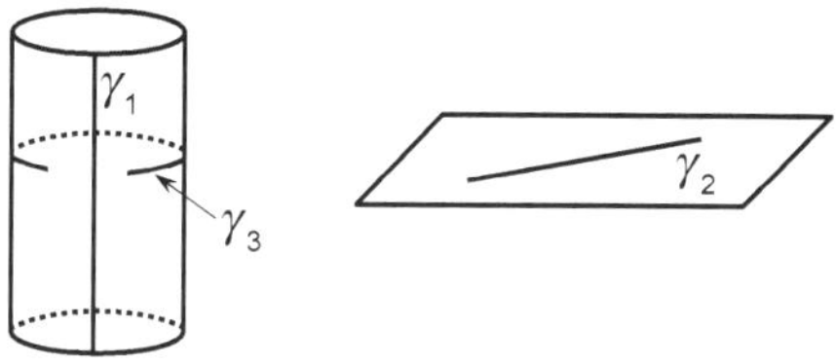

Thus, the extrinsic curvature represents the amount by which "curves which insist on remaining in S, but otherwise try to be as geodesic as possible, are forced to curve in order to remain in S". Why is the right side of (118) normal to S? Physical analogy: consider a point particle, free to move in R^3 except that it is constrained to lie in some surface S. Then the surface S will find it necessary to exert a force on the particle to keep it in S, and this force will be normal to S.

Exercise 190. Find the extrinsic curvature for the plane and cylinder examples, and verify explicitly Eqn. (118).
Example 75. A curve, considered as a submanifold, is a geodesic if and only if its extrinsic curvature vanishes.
Example 76. For co-dimension zero, the extrinsic curvature always vanishes.
Exercise 191. Let S be a manifold with metric h'_{ab}. Consider R^m with its natural positive-definite metric. Set $M = S \times R^m$, and let g_{ab} be the metric on M such that the induced metric on each submanifold S and each submanifold

R^m is the given one. (This construction is called the *product* of manifolds-with-metric.) Consider the submanifold (a copy of S) consisting of pairs (p, x) with p in S and $x = 0$ (in R^m). Prove that its extrinsic curvature is zero. (Conclusion: The intrinsic geometry of a submanifold allows no prediction about what its extrinsic curvature will be.)

Exercise 192. Find an example of a metric submanifold whose intrinsic metric has vanishing Riemann tensor, but whose extrinsic curvature is nonzero.

Exercise 193. Find all 2-dimensional submanifolds of R^3 (with usual metric) having vanishing extrinsic curvature.

Exercise 194. Find the extrinsic curvature of S^{n-1} in R^n (usual embedding, usual metric.).

Exercise 195. Find a manifold-with-metric such that no submanifold of co-dimensional one has vanishing extrinsic curvature.

Exercise 196. Let S be a metric submanifold of M, T a metric submanifold of S. Let S have vanishing extrinsic curvature in M, and T vanishing extrinsic curvature in S. Prove that T vanishing extrinsic curvature in M.

Exercise 197. Let M be a manifold-with-metric, S a (not necessarily metric) submanifold. How much of this extrinsic curvature business goes through?

33. The Gauss-Codazzi Equations

Let S be a metric submanifold of M (with g_{ab}). There exists a set of two equations relating the geometry of M, the geometry of S, and the extrinsic curvature of S. These are called the Gauss-Codazzi equations. They turn out to be useful in numerous applications. It is perhaps not surprising that such equations should exist. Consider a two-dimensional submanifold of R^3 (with the usual metric). Intuitively, if the surface is very "wiggly" in R^3, then it would tend to have a large intrinsic Riemann tensor, while a relatively "flat" S would have a small Riemann tensor.

The first Gauss-Codazzi equation arises from the following fact from Sect. 31: there is a natural, one-to-one, correspondence between tensor fields on the manifold S and tensor fields on the submanifold S all of whose indices are tangent to S. Furthermore, the mapping from tensor fields on the submanifold S all of whose indices are tangent to S to tensor fields on the submanifold S all of whose indices are tangent to S defined by

$$h^b{}_m \cdots h^c{}_n h_d{}^p \cdots h_e{}^q\, D_a \alpha^{m\ldots n}{}_{p\ldots r} \tag{119}$$

corresponds precisely to the mapping from tensor fields on the manifold S to tensor fields on the manifold S defined by the derivative operator on S (associated with the metric h'_{ab} of S). Thus, the Riemann tensor of the manifold S can be obtained by a calculation in M. Specifically, if k_a is a vector field on the sub manifold S which is tangent to S, then

$$\frac{1}{2}\mathcal{R}_{abc}{}^d k_d = h_{[a}{}^m h_{b]}{}^n h_c{}^p \nabla_m (h_n{}^r h_p{}^s \nabla_r k_s) \tag{120}$$

where $\mathcal{R}_{abc}{}^d$ is the Riemann tensor of S. The idea is to expand the right side of (120):

$$\frac{1}{2}\mathcal{R}_{abc}{}^d k_d = h_{[a}{}^m h_{b]}{}^n h_c{}^p [(\nabla_m h_n{}^r) h_p{}^s \nabla_r k_s) \tag{121}$$
$$+ h_n{}^r (\nabla_m h_p{}^s) \nabla_r k_s + h_n{}^r h_p{}^s \nabla_m \nabla_r k_s]$$

The first term on the right in (121) vanishes, for $h_{[a}{}^m h_{b]}{}^n \nabla_n h_n{}^r = \pi_{[ab]}{}^r = 0$. The last two terms in (121) give

$$
\begin{aligned}
\frac{1}{2}\mathcal{R}_{abc}{}^{d}k_d &= h_{[a}{}^{m}h_{b]}{}^{r}h_c{}^{p}(\nabla_m h_p{}^{s})\nabla_r k_s + h_{[a}{}^{m}h_{b]}{}^{r}h_c{}^{s}\nabla_m\nabla_r k_b \\
&= h_{[b}{}^{r}\pi_{a]c}{}^{s}\nabla_r k_s + h_a{}^{m}h_b{}^{r}h_c{}^{s}\nabla_{[m}\nabla_{r]}k_s \\
&= h_{[b}{}^{r}\pi_{a]c}{}^{s}\nabla_r k_s + \frac{1}{2}h_a{}^{m}h_b{}^{r}h_c{}^{s}R_{mrs}{}^{d}k_d
\end{aligned}
\tag{122}
$$

were, for the first step, we have used $h_a{}^{m}h_m{}^{b} = h_a{}^{b}$ (repeatedly), for the second step, we have used the definition of $\pi_{ab}{}^{c}$, for the third step, we have used the definition of the Riemann tensor, $R_{abc}{}^{d}$, of the manifold M. Finally, we modify the first term on the far right side of (122) as follows:

$$
\begin{aligned}
h_{[b}{}^{r}\pi_{a]c}{}^{s}\nabla_r k_s &= h_{[b}{}^{r}\pi_{a]c}{}^{s}\nabla_r(h_s{}^{m}k_s) = h_{[b}{}^{r}\pi_{a]c}{}^{s}k_m\nabla_r h_s{}^{m} \\
&+ h_{[b}{}^{r}\pi_{a]c}{}^{s}h_s{}^{m}\nabla_r k_m = h_{[b}{}^{r}\pi_{a]c}{}^{s}k_m\nabla_r h_s{}^{m} = \pi_{c[a}{}^{s}\pi_{b]ms}k^m
\end{aligned}
$$

where, in the third step, we have used the fact that $\pi_{ac}{}^{s}h_s{}^{m} = 0$ (a normal index contracted with a tangent index). Thus, (122) becomes

$$
\frac{1}{2}\mathcal{R}_{abc}{}^{d}k_d = \pi_{c[a}{}^{s}\pi_{b]ms}k^m + \frac{1}{2}h_a{}^{m}h_b{}^{r}h_c{}^{s}R_{mrs}{}^{d}k_d
$$

Using the fact that k_c is arbitrary (tangent to S), we obtain the first *Gauss-Codazzi equation*:

$$
\mathcal{R}_{abcd} = 2\pi_{c[a}{}^{m}\pi_{b]dm} + h_a{}^{m}h_b{}^{n}h_c{}^{p}h_d{}^{q}R_{mnpq}
\tag{123}
$$

This equation states that the Riemann tensor of the sub manifold S (considered as a manifold-with-metric in its own right) is equal to the Riemann tensor of M, with its indices all projected tangent to S, plus a term involving the extrinsic curvature. Intuitively, the "intrinsic wiggliness of M" (described by the Riemann tensor of M) and the "wiggliness of the embedding of S in M" (described by the extrinsic curvature) combine to give the "intrinsic wiggliness of S" (described by the Riemann tensor of S).

To obtain the second Gauss-Codazzi equation, we proceed as follows:

$$
\begin{aligned}
h_{[a}{}^{m}h_{b]}{}^{n}h_c{}^{p}k_q{}^{d}\nabla_m\pi_{np}{}^{q} &= h_{[a}{}^{m}h_{b]}{}^{n}h_c{}^{p}h_q{}^{d}\nabla_m[h_n{}^{r}h_p{}^{s}k^{qt}\nabla_r h_{st}] \\
&= h_{[a}{}^{m}h_{b]}{}^{n}h_c{}^{p}h_q{}^{d}[(\nabla_m h_n{}^{r}h_p{}^{s}k^{qt}\nabla_r h_{st} \\
&+ h_n{}^{r}(\nabla_m h_p{}^{s})k^{qt}\nabla_r h_{st} + h_n{}^{r}h_p{}^{s}(\nabla_m k^{qt})\nabla_r h_{st} + h_n{}^{r}h_p{}^{s}k^{qt}\nabla_m\nabla_r h_{st}]
\end{aligned}
\tag{124}
$$

where, in the first step, we have used the definition of the extrinsic curvature, and, in the second, we have expanded the derivative of the product. Now, each of the first three terms in the last expression in (124) vanishes. The first term vanishes because $h_{[a}{}^{m}h_{b]}{}^{n}\nabla_m h_n{}^r = \pi_{[ab]}{}^r = 0$. The second term vanishes for a more subtle reason. Since the indices "m" and "p" of $(\nabla_m h_p{}^s)$ are being projected tangent to S, the index "s" here is normal to S. Thus, for the factor $\nabla_r h_{st}$, the "s" is projected normal to S, and also "t" is being projected normal to S (because of the k^{qt}), But, if both indices "s" and "t" of $\nabla_r h_{st}$ are projected normal to S, we obtain zero. (See the calculation preceding (115).) The third term vanishes by an argument similar to that for the second. Consider the factor $(\nabla_m k^{qt})$. Here, the "q" is being projected normal to S by $(k_q{}^d)$, and the "m" is being projected tangent to S (by $h_a{}^m$). Hence, the "t" in $(\nabla_m k^{qt})$ is tangent to S. But now consider the factor $\nabla_r h_{st}$. Both "s" and "t" are being projected tangent to S, so we get zero. Thus, the right side of (124) becomes simply

$$h_{[a}{}^{m}h_{b]}{}^{n}h_c{}^s k^{dt}\nabla_m\nabla_n h_{st} = h_a{}^m h_b{}^n h_c{}^s k^{dt}\nabla_{[m}\nabla_{n]}h_{st}$$
$$= h_a{}^m h_b{}^n h_c{}^s k^{dt}\frac{1}{2}\left[R_{mns}{}^p h_{pt} + R_{mnt}{}^p h_{sp}\right]$$

So, we have

$$h_{[a}{}^{m}h_{b]}{}^{n}h_c{}^p k_d{}^q\nabla_m\pi_{npq} = -\frac{1}{2}h_a{}^m h_b{}^n h_c{}^p k_d{}^q R_{mnpq} \qquad (125)$$

This is the second *Gauss-Codazzi equation.* It looks at the Riemann tensor of M with three indices projected tangent to S and one normal, and expresses it in terms of the rate of change of the extrinsic curvature of S (properly projected). Note that the Riemann tensor of S does not appear above. Thus, curvature of M of a certain sort reflects itself in a derivative of the extrinsic curvature. (Why the projections on the left in (125)? Note that, if any of the h's or the k is replaced by the other, then the result can easily be expressed algebraically in terms of the extrinsic curvature by a single integration by parts.)

Although the calculations above look rather formidable, time spend staring at them is well worthwhile. It is usually easier to derive the Gauss-Codazzi equations when needed than to go down two flights of stairs to the library to look them up. It should now be clear why we spent so much time at the beginning establishing an index notation!

We shall give examples of the use of these equations shortly.

34. Metric Submanifolds of Co-dimension One

The discussion of the previous sections has been quite general: for a $k-$dimensional submanifold of an $n-$dimensional manifold. Easily the most important case for applications is that in which $k = n - 1$, i.e., the case of co-dimension one. Things can be simplified slightly in this case. We now carry out this simplification.

The crucial consequence of co-dimension one is the following: the space of vectors normal to S at each point of S is one-dimensional. Hence, it consists of multiples of some vector ξ^a. Furthermore, we can normalize ξ^a by

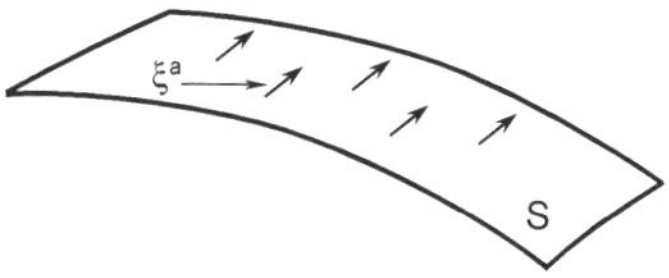

$\xi^a \xi_a = \pm 1$. (The choice of sign is forced by the signature of g_{ab} and by S. We shall carry both signs through the calculations.) Thus, the choice of ξ^a is determined uniquely up to sign. (Of course, reversing the sign of ξ^a leaves the sign of $\xi^a \xi_a$ unchanged.) Since the space of normal vectors is one-dimensional, it follows that, if "a" of $\alpha_{ab...c}$ is normal to S, then $\alpha_{ab...c} = \xi^a \beta_{b...c}$. Thus,

$$h_{ab} = g_{ab} \mp \xi_a \xi_b \qquad k_{ab} = \pm \xi_a \xi_b \tag{126}$$

Set

$$\pi_{abc} = \mp \pi_{ab} \xi_c \tag{127}$$

Thus, π_{ab} is symmetric, and both its indices are tangent to S. The extrinsic curvature, a tensor field with three indices, two tangent and one normal to S, reduces to π_{ab}. In co-dimension one, this π_{ab} is called *extrinsic curvature*. To express π_{ab} directly in terms of ξ^a, we proceed as follows:

$$\begin{aligned} \pi_{ab} = -\xi^c \pi_{abc} &= -h_b{}^m \xi^c D_a h_{mc} = -h_b{}^m \left[D_a \left(\xi^c h_{mc} \right) - h_{mc} D_a \xi^c \right] \\ &= h_b{}^m h_{mc} D_a \xi^c = h_b{}^m D_a \xi^m \end{aligned} \tag{128}$$

Eqn. (128) allows an alternative interpretation of the extrinsic curvature in the case of co-dimension one. Let $\tilde{\xi}^a$ be an extension of ξ^a, with $\tilde{\xi}^a \tilde{\xi}_a =$

± 1, and set $\tilde{h}_{ab} = g_{ab} \mp \tilde{\xi}_a \tilde{\xi}_b$ so $\tilde{h}_{ab}$ is an extension of h_{ab}. Next, note that.

$$\tilde{\xi}^a \mathscr{L}_{\tilde{\xi}} \tilde{h}_{ab} = \mathscr{L}_{\tilde{\xi}}(\tilde{h}_{ab}\tilde{\xi}^a) - \tilde{h}_{ab}\mathscr{L}_{\tilde{\xi}}\tilde{\xi}^a = 0 - 0 = 0$$

so $\mathscr{L}_{\tilde{\xi}}\tilde{h}_{ab}$ is tangent to S. Thus,

$$\begin{aligned}
\mathscr{L}_{\tilde{\xi}}\tilde{h}_{ab} &= \tilde{h}_a{}^m \tilde{h}_b{}^n \mathscr{L}_{\tilde{\xi}}\tilde{h}_{mn} = \tilde{h}_a{}^m \tilde{h}_b{}^n \mathscr{L}_{\tilde{\xi}}(g_{mn} \mp \tilde{\xi}_m\tilde{\xi}_n) \\
&= \tilde{h}_a{}^m \tilde{h}_b{}^n [\mathscr{L}_{\tilde{\xi}} g_{mn} \mp \tilde{\xi}_m \mathscr{L}_{\tilde{\xi}}\tilde{\xi}_n \mp \tilde{\xi}_n \mathscr{L}_{\tilde{\xi}}\tilde{\xi}_n] = \tilde{h}_a{}^m \tilde{h}_b{}^n \mathscr{L}_{\tilde{\xi}} g_{mn} \\
&= \tilde{h}_a{}^m \tilde{h}_b{}^n [\tilde{\xi}^p \nabla_p g_{mn} + g_{mp}\nabla_n \tilde{xi}^p + g_{pn}\nabla_m \tilde{xi}^p] = 2\tilde{h}_a{}^m \tilde{h}_b{}^n \nabla_{(m}\tilde{\xi}_{n)} \\
&= 2\pi_{ab}
\end{aligned}$$

The extrinsic curvature thus represents the "Lie derivative of h_{ab} in the direction normal to S". The reason why the extrinsic curvature of the plane and cylinder (in R^s) are what they are is clear from this point of view.

Finally, we substitute (127) into Gauss-Codazzi equations. The resulting (123) is obvious:

$$\mathcal{R}_{abcd} = \pm 2\pi_{c[a}\pi_{b]d} + h_a{}^m h_b{}^n h_c{}^p h_d{}^q R_{mnpq} \tag{129}$$

For (125), note that we might as well contract this equation with ξ^d (for the "d" index is normal to S). Substituting (127) into (125), and nothing that $\xi^q D_m \xi_q = 0$, we have

$$h_{[a}{}^m h_{b]}{}^n h_c{}^p \nabla_m \pi_{np} = \mp\frac{1}{2} h_a{}^m h_b{}^n h_c{}^p \xi^d R_{abcd} \tag{130}$$

These are the *Gauss-Codazzi equations* for co-dimension one.

Example 77. Let $M = R^3$, and let S be the S^2 in M given by $(x^1)^2 + (x^2)^2 + (x^3)^2 = 1$. Let g_{ab} be the metric on M whose components, in the natural chart, are $g_{11} = g_{22} = g_{33} = 1$, others zero. Then S is a metric submanifold. Denote by x_a the vector field on M whose components, in the natural chart, are (x_1, x_2, x_3). Then $\nabla_a x_b = g_{ab}$. On S, x_a is normal to S. From (128)

$$\pi_{ab} = h_b{}^m D_a x_m = h_{ab}$$

We now consider the Gauss-Codazzi equations. Note that R_{abcd}, the Riemann tensor of M, vanishes. Eqn. (130) just reduces to an identity. From (129), we have

$$\mathcal{R}_{abcd} = 2h_{c[a}h_{b]d}$$

This is the expression for the Riemann tensor of the (metric) S^2. Hence,$\mathcal{R} =$ 2

Exercise 198. Adapt the example above to find the extrinsic curvature and Riemann tensor for the plane in R^3.

Exercise 199. Adapt the example above to find the extrinsic curvature and Riemann tensor for the cylinder in R^3.
Exercise 200. Adapt the example above to find the extrinsic curvature of S^n in R^{n+1}.
Exercise 201. Let $M = R^4$, with metric, in the natural chart, $g_{11} = g_{22} = g_{33} = 1, g_{44} = -1$. others zero. Let S be the submanifold consisting of points with $(x^1)^2 + (x^2)^2 + (x^3)^2 - (x^4)^2 = -1$. (This is called the unit hyperboloid.) Show that S is a metric submanifold, and find its extrinsic curvature and Riemann tensor.
Exercise 202. Show that the Gauss-Codazzi equations are vacuous for submanifolds of dimension one; same for co-dimension zero.
Exercise 203. Show that the result of substituting the right side of (129) into the Bianchi identity on $\mathcal{R}_{abcd}$ is an identity.
Exercise 204. Convince yourself that $h_a{}^m h_c{}^n \xi^b \xi^d R_{mbnd}$ cannot be expressed in terms of the extrinsic and its D–derivatives. (Conclusion: There are no additional Gauss-Codazzi equations.)
Exercise 205. Using the interpretation on p. 116, show that, if M is considered as a sub manifold of $M \times R^1$ (a product of manifolds with metric), then the extrinsic curvature of M vanishes.
Exercise 206. using the intuitive interpretation of the Riemann tensor as giving the "change in a vector under parallel transport about an infinitesimal curve", and of the extrinsic curvature as "the date of change of the tangent plane to a submanifold", intuit the validity of the Gauss-Codazzi equations.

35. Orientation

Let M be an n–dimensional manifold, and let p be a point of M. We introduce some properties of antisymmetric tensors at p.

Let $\mu_{a...c}$ be an antisymmetric tensor at p of rank n. Then, introducing a chart containing p, we consider the components of $\mu_{a...c}, \tilde{\mu}_{i...j}$. Since $\mu_{a...c}$ is antisymmetric, the only nonzero components are those for which $i \ldots j$ are all different. Hence, $\mu_{a...c}$ is completely determined by the numerical value of the single component $\tilde{\mu}_{1...n}$. Thus, if $\mu_{a...c}$ and $\nu_{a...c}$ are nonzero number α such that $\mu_{a...c} = \alpha\nu_{a...c}$.

Now consider the collection of all nonzero, antisymmetric, rank n, covariant tensors at p. We regard two such tensors as equivalent if the proportionality factor between them is positive. Evidently, there are precisely two equivalence classes. (The vector space of such tensors at p is one-dimensional. We remove the origin, i.e., we exclude the zero tensor. There remain two "pieces", which are precisely the equivalence classes.) A choice of one of these equivalence classes at p is called an *orientation* of the cotangent space at p. Thus, there are precisely two orientations of the cotangent space at p.

The same remarks apply, of course, to contravariant tensors. Similarly, we define an orientation of the tangent space of p. In fact, an orientation of the cotangent space at p defines an orientation of the tangent space at p and vise versa. We shall prove slightly more: if $\epsilon_{a...c}$ is nonzero, rank n, and antisymmetric, then there is precisely one nonzero, rank n, antisymmetric $\epsilon^{a...c}$ (all at p) such that

$$\epsilon^{a...c}\epsilon_{a...c} = n! \tag{131}$$

(The choice of the $n!$ is for convenience. If, with respect to a chart, $\epsilon_{1...n} = 1$. then, with respect to that same chart, $\epsilon^{1...n} = 1$). To prove this, let $\mu^{a...c}$ be nonzero, and set $\alpha = \mu^{a...c}\epsilon_{a...c}$. Note (e.g., using a chart) that $\alpha \neq 0$. Hence, $\epsilon^{a...c} = n!\,\alpha^{-1}\mu^{a...c}$ is the (obviously unique) solution of (131). The $\epsilon^{a...c}$ of (131) is called the *inverse* of $\epsilon_{a...c}$. Noting that, if two nonzero $\epsilon_{a...c}$'s differ by a positive factor, then so do their inverses, we obtain a correspondence between the (two) orientations of the cotangent space and those of the tangent space.

The sense in which $\epsilon^{a...c}$ is the "inverse" of $\epsilon_{a...c}$ is the following. Let $\mu_{a...c}$ be antisymmetric. Then, since μ is a multiple of $\epsilon_{a\cdots c}$, the left side of

$$\frac{1}{n!}\mu_{m\cdots n}\epsilon^{m\cdots n}\epsilon_{a...b} = \mu_{a\cdots c} \tag{132}$$

is a manifold of the right. Contracting with $\epsilon^{a\cdots c}$, we see that (132) holds.

Roughly speaking, an orientation of a manifold M consists of a continuous assignment of an orientation to the tangent space of each point of M. More precisely, consider two tensor fields, $\mu_{a...c}$ and $\nu_{a...c}$, on M, which are antisymmetric and of rank n. Suppose neither vanishes anywhere. Hence, $\mu_{a...c} = \alpha\nu_{a...c}$, where α is a scalar field on M. (Exercise 207. Prove smooth.) These two fields will be regarded as equivalent if α is everywhere positive. (Note that, in any case, α vanishes nowhere.) An equivalence class of such fields defines an *orientation* of M. As we shall see in a moment, a manifold M may have no orientation. Suppose that M is connected. Then any scalar field on M which vanishes nowhere is either everywhere positive or everywhere negative. (Exercise 208. Prove this statement.) In this case, therefore, if there is any orientation on M, there are precisely two.

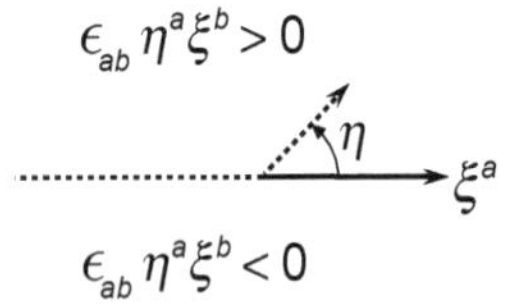

Finally, we give a geometrical picture of orientation in the case of dimension two. Let ϵ_{ab} be a nonzero skew tensor at a point p of a 2–dimensional manifold. Fix a vector ξ^a at p, and consider the quantity $\epsilon_{ab}\eta^a\xi^b$ as η^a varies through vectors at p. This quantity is zero for $\eta^a = \xi^a$, and, as η^a "rotates" away from ξ^a, becomes either positive or negative. It is again zero for $\eta^a = -\xi^a$. Hence, the direction in which η^a must rotate from ξ^a to make the quantity above positive defines a "direction of rotation" from ξ^a. Thus, an orientation on a 2–dimensional manifold M represents a "direction of rotation" in the tangent space at each point of M.

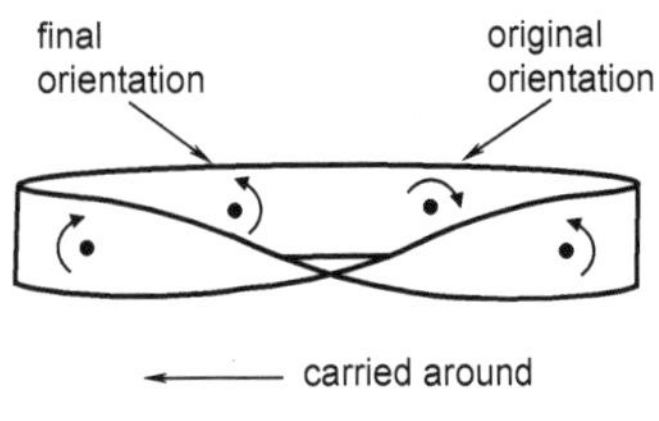

The question of whether a manifold has an orientation is the question of whether or not there exists a contines choice of rotation direction at each point. It is perhaps not surprising that a Mobius strip has no orientations.

36. Integrals

We now introduce the notion of an integral over a manifold. We begin with integrals in R^n. Let U be an open subset of R^n, and let $f(x)$ (where $x = (x^1, \ldots, x^n)$) be a function on U. Then, of course, the Riemann integral,

$$\int_U f(x)\, dx^1 \cdots dx^n \tag{133}$$

is well defined. (When we write an integral, the assumption that the integral converges absolutely is automatically in force. if an integral does not converge, it does not converge, and the discussion ends.) Suppose we now perform a coordinate transformation, $x'^1 = x'^1(x), \ldots, x'^n(x) = x'^n(x)$. Then, from elementary calculus,

$$\int_U f(x)\, dx^1 \cdots dx^n = \int_U f(x(x')) \left|\frac{\partial x}{\partial x'}\right| dx'^1 \cdots dx'^n \tag{134}$$

where $|\frac{\partial x}{\partial x'}|$ is the Jacobian, i.e., the absolute value of the determinant of the $n \times n$ matrix $\partial x^i/\partial x'^j$ $(i, j = 1, \ldots, n)$.

It is clear from (134) at there is no immediate definition of the integral of a function f over an open region U of a manifold M. The most obvious definition would be as follows: choose a chart containing U, and define $\int_U f$ to be (133). But this is no good, for, by (134), the result depends on the choice of chart. Of course, any reasonable definition of an integral depends only on the quantity being integrated and the region of M over which the integration takes place.

How do we arrange things so we get a reasonable notion of integration in a manifold? Let $f_{a \ldots c}$ be an n–form on the (n–dimensional) manifold M, and let U be an open subset of M sufficiently small that there exists a chart on U M containing U. Choose two such charts, with coordinates x^i and x'^i $(i = 1, \ldots, n)$. Then the components of $f_{a \ldots c}$ in these charts are related by

$$f'_{i \cdots j} = \Sigma f_{k \cdots l} \frac{\partial x^k}{\partial x'^i} \cdots \frac{\partial x^k}{\partial x'^j} \tag{135}$$

From (135), it follows that

$$f'_{i\cdots j} = f_{i\cdots j}\det(\frac{\partial x^k}{\partial x'^l}) \tag{136}$$

(Proof: The left side of (136) is certainly proportional to the right, by Sect. 35. To prove equality, contract both sides with $\epsilon^{i\cdots j}$, use (132), and the fact that

$$\det K_a{}^b = \frac{1}{n!} K_a{}^{b\cdots} K_d{}^c \epsilon^{a\cdots d} \epsilon_{b\cdots c} \tag{137}$$

for a tensor $K_a{}^b$ where $\epsilon^{a\ldots d}$ and $\epsilon_{b\ldots c}$ are inverses.)

We are now almost prepared to define integration on a manifold. The only remaining difficulty is the presence of an absolute value (in the Jacobian) in (134). To correct this, we only admit coordinates x^i $(i = 1, \ldots, n)$ in which $\epsilon_{1\cdots n} > 0$, where $\epsilon_{a\cdots b}$ represents an orientation on M. Let $f_{a\ldots c}$ be an $n-$form on the oriented ($n-$dimensional) manifold M. Choose a chart compatible (in the sense above) with the orientation of M, and define the left side of

$$\int_U f_{a\cdots c} ds^{a\cdots c} = n! \int_U f_{1\cdots n}\, dx^1 \cdots dx^n \tag{138}$$

by the right side. The result is independent of choice of chart, for, by (134) and (136),

$$\int_U f_{1\cdots n} dx^1 \cdots dx^n = \int_U f_{1\cdots n} \left|\frac{\partial x}{\partial x'}\right| dx'_1 \cdots dx'^n = \int_U f'_{1\cdots n}\, dx'^1 \cdots dx'^n$$

Thus, integration should be performed on an oriented manifold, and the proper thing to integrate is an $n-$form. (These are not all that much different from functions. Each forms are one-dimensional vector space at each point. The nice thing about forms is that their components have the correct behavior, e.g., (134), under coordinate transformations.

As in elementary calculus, we define integrals (of $n-$forms on an oriented manifold) over arbitrary open regions U by the requirement that, if U is contained in a chart, the integral is given by (138), and for U and V disjoint,

$$\int_U + \int_V = \int_{U\cup V} \tag{139}$$

Finally, note that, if $\mu_{a\cdots c}$ and $\nu_{a\cdots c}$ are $n-$forms, s a constant, then

$$\int_U (\mu_{a\cdots c} + s\nu_{a\cdots c}) ds^{a\ldots c} = \int_U \mu_{a\cdots c} ds^{a\cdots c} + s \int_U \nu_{a\cdots c} ds^{a\cdots c} \tag{140}$$

Integration on a a manifold is not much different from integration in elementary calculus. The only difference is that one integrates an $n-$form instead of a function.

The thing one integrates over a $k$dimensional submanifold of M is a $k$form. Let $\psi : S \to M$ be a submanifold of M, and suppose that S is oriented (i.e., at each point of $\psi[S]$ an equivalence class of antisymmetric, rank k tensors, all of whose indices are tangent to S, is specified). The left side of

$$\int_{\psi[U]} \omega_{a\cdots b}\, ds^{a\cdots b} = \int_U (\underset{\leftarrow}{\psi}\, \omega_{a\cdots b})\, ds^{a\cdots b} \qquad (141)$$

is defined by the right, where U is an open subset of S, and $\omega_{a\cdots b}$ is a k–form on M.

37. Stokes' Theorem

Recall the equation of the fundamental theorem of calculus

$$\int_a^b f'(x)\,dx = f(b) - f(a) \tag{142}$$

There is a generalization of this result to manifolds. The integral over the real line in (142) is replaced by an integral over a manifold, the function f in (142) by a form, and the derivative in (142) by the exterior derivative. Making these replacements, we have

$$\int_U D(f)_{ma\cdots b}\, ds^{ma\cdots b} = \int_{\partial U} f_{a\cdots b}\, ds^{a\cdots b} \tag{143}$$

where $f_{a\cdots b}$ is an $(n-1)$–form, U is an open subset of M, and ∂U is the "boundary" of U. Accompanied by an appropriate collection of definitions and qualifications, Eqn. (143) is indeed true. It is this result we wish to establish in the present section.

Let M be an n–dimensional manifold, U and open subset of M. The *boundary* of U, ∂U, is the collection of all points p of M such that p is not in U, but every chart containing p intersects U. This definition corresponds, of course, to the intuitive notion of the boundary of an open set. (In the figure on the right, p is in the boundary of U, but neither q nor r are in ∂U.)

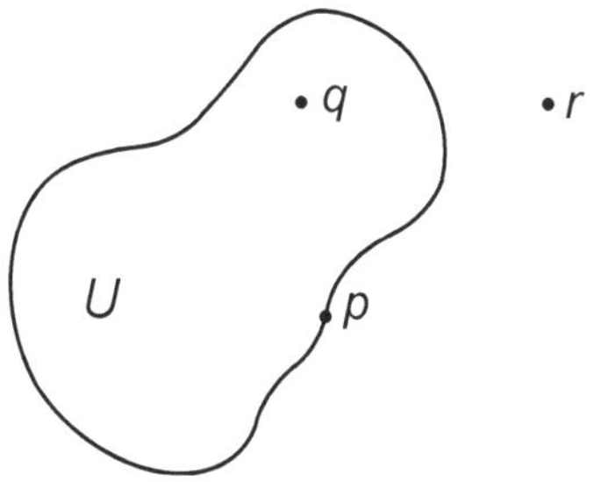

In order that the right side of (143) make sense, it is necessary that ∂U be an $(n-1)$–dimensional submanifold of M.

We next deal with the orientation in (143). Let M be oriented. Thus, at each point of M, we have selected one of the two classes of equivalent (nonzero) n–forms. Now the integral on the left in (143) makes sense. What orientation should we choose for ∂U on the right? Fix a point p of ∂U. Then the covariant vectors (in M) at p which are normal to the $(n-1)$–dimensional sub manifold M form a one-dimensional subspace of the cotangent space of p. This one-dimensional vector space is divided into two pieces by its zero element. We want to regard one of these pieces as consisting of "outgoing normals" (i.e., heading away from U), and the other as consisting of "ingoing normals". Such a distinction would, evidently, be impossible of U "met ∂U on both sides of ∂U". (E.g., as in p in the figure below on the right rather than q.) We are thus led to the following definition: ∂U will be said to *border* the open set U if ∂U is an $(n-1)$–dimensional submanifold and if, for each point p of ∂U, there is a chart containing p such that the intersection of this chart with U consists precisely of the points of the chart with $x^n < 0$. Thus, in the first figure above, ∂U, while in the second it does not.

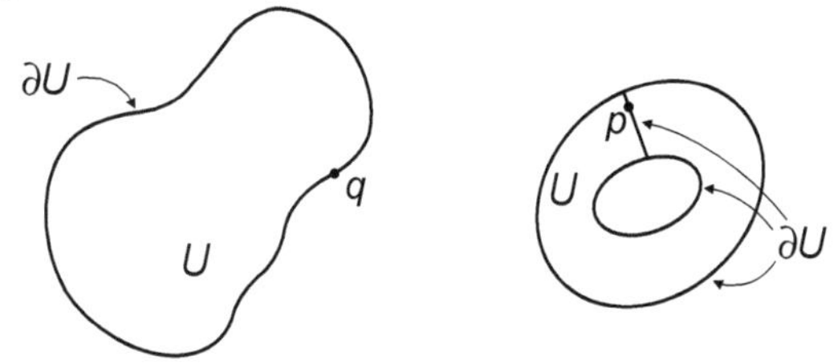

A covariant vector μ_a normal to ∂U at a point p of ∂U will be said to be an *outgoing* normal if, with respect to the chart above, the nth component of μ_a is positive. Evidently, ν_a is another outgoing normal if and only if ν_a is a positive multiple of μ_a.

Finally, we are ready to induce an orientation on ∂U from that of M. Let p be a point of ∂U. and let $\lambda^{a\cdots c}$ be antisymmetric of rank n, and in the equivalence class defining the orientation of M. Let μ_a be an outgoing normal to ∂U. Then $\mu_a\lambda^{ab\cdots c}$ is antisymmetric, of rank $(n-1)$, and with all indices tangent to ∂U. This tensor thus defines an orientation of ∂U. We call this orientation the induced orientation on ∂U. Note that, if we reverse the orientation originally assigned to M, we also reverse the induced orientation on ∂U.

Thus, if we assume that ∂U borders U and use the induced orientation on the right in (143), everything in that equation is well-defined. Is the equation now true in general? Unfortunately, the answer is no. We still require one further condition. In order to see why this condition is necessary, we begin the proof of (143).

Let U be the region of R^n consisting of points $(x^1, \ldots, x^n)$ with $0 < x^1 < 1, \ldots, 0 < x^n < 1$. Let $f(x)$ be a function on R^n. Then

$$\int_U \frac{\partial f}{\partial x^1}\, dx^1 \cdots dx^n = \int_{x^1=1} f\, dx^2 \cdots dx^n - \int_{x^1=0} f\, dx^2 \cdots dx^n \qquad (144)$$

where the two integrals on the right are surface integrals over appropriate faces on the cube U. Next, let $f_{i\ldots j}(x)$ be the components of some $(n -$

1)–form on R^n. Then the components of the exterior derivative of this form are

$$\frac{1}{n}\left[\frac{\partial}{\partial x^1} f_{2\cdots n} + \epsilon \frac{\partial}{\partial x^2} f_{3\cdots n1} + \cdots + \epsilon \frac{\partial}{\partial x^n} f_{1\cdots(n-1)}\right] = D(f)_{1\cdots n} \qquad (145)$$

$$\epsilon = \begin{cases} 1 & n \text{ odd} \\ -1 & n \text{ even} \end{cases}$$

From (144) and (145), we see that Eqn. (143) is valid for this U provided we take for ∂U on the right the union of the (plane) faces of U (leaving off the corners and edges, etc. We know from elementary calculus that these are irrelevant. Strictly speaking, (143) is not meaningful for our U because the actual ∂U is not a sub manifold: it has corners.)

From these remarks, the following statement should be clear: Eqn. (143) is valid if there is a chart containing U and ∂U such that the image of $U_\cup \partial U$ in R^n is closed and bounded.

The additional condition we require is that U does not "go off to infinity in M" so that "some of the surface integral on the right in (143) is not included in (143) because the appropriate boundary is off at infinity". In other words, we must ensure that none of the flux escapes to infinity: that all is properly taken into account by the surface integral on the right in (143). For the U on the right, for example, we could not guarantee that (143) is valid.

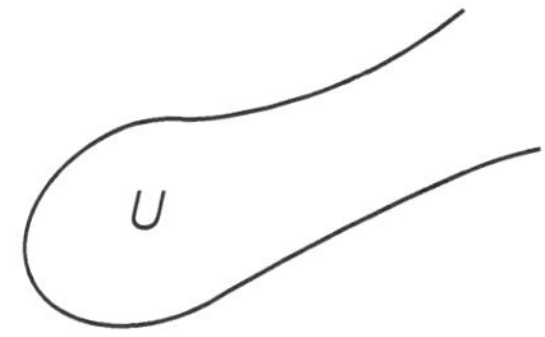

A subset C of M will be said to be chart-compact if there is a chart containing C such that the image of C in R^n is closed and bounded. A subset C of M is said to be *compact* if C can be written as a union of chart-compact sets. (As usual, we have merely defined a standard topological notion on M, but using chart instead of topological notion.)

The following statement is true:

Theorem 22. Let U be an open subset of M such that ∂U borders U, and such that $U_\cup \partial U$ is compact. Let M be oriented. Then, using the induced orientation on ∂U, Eqn. (143) holds. We have not, of course, proven Theorem 22. It should be clear from the discussion above, however, that the techniques of elementary calculus together with a certain amount of work will suffice to write out a full proof.

Since integrals over sub manifold reduce to integrals over manifolds, Theorem 22 is also true for submanifolds. These results are called *Stokes' Theorem.* Essentially every situation in practice in which an integral "cancels" a derivative is a special case of Stokes' theorem.

Example 78. Let S be a one-dimensional submanifold of M, so S consists of a curve. Let p and q consist of the endpoints of this curve. Then Stokes' theorem becomes

$$\int_S D(\alpha)_a \, ds^a = \alpha(q) - \alpha(p)$$

for any scalar field α.

Example 79. Let U consist of the points of R^2 with $(x^1)^2 + (x^2)^2 < 1$. Then ∂U consists of points with $(x^1)^2 = (x^2)^2 = 1$. In this case, ∂U borders U.

Example 80. Let U consist of points of R^2 with $(x^1)^2 + (x^2)^2 \neq 1$. Then ∂U is as above, but now ∂U does not border U.

Exercise 207. Find the orientations of S^2, of R^2.

Exercise 208. Using Example 78, prove that a scalar field on a connected manifold, having vanishing gradient, is constant.

Exercise 209. Use Theorem 22 to prove that the exterior derivative, applied twice in succession, gives zero. (Hint: the left side of (143) remains the same if U is replaced by any other open set with the same boundary as U.)

Exercise 210. Verify (143) explicitly in R^2 with the U of Example 79, and with f_a with components $(x^2, -x^1)$. That R^2 is not compact.

Exercise 211. Show that the manifold S^2 itself is compact. That R^2 is not compact.

Exercise 212. Consider the geometrical interpretation of orientation (for a two-dimensional manifold) in Sect. 35. Obtain a similar interpretation for a one-dimensional submanifold. Using these interpretations, describe the operation of obtaining the induced orientation on the boundary of an open subset of a two-dimensional manifold.

Exercise 213. Derive (142) from (143).

38. Integrals: The Metric Case

On a manifold with metric, there is an alternative way of writing integrals. This formation is often the most convenient and suggestive.

Let M be an oriented $n-$dimensional manifold with metric g_{ab}. Let $\omega_{a\cdots c}$ be an $n-$form at a point p of M, where this $n-$form is consistent with the orientation of M. Then, since g_{ab} is invertible, the quantity

$$\omega_{a\cdots c}\omega^{a\cdots c} = \omega_{a\cdots c}(g^{am}\cdots g^{cn}\,\omega_{m\cdots n}) \tag{146}$$

is nonzero. If $\omega_{a\cdots c}$ in (146) is replaced by $\alpha\omega_{a\cdots c}$, where α is a number, then the expression (146) is multiplied by α^2. Thus, there exists precisely one tensor $\epsilon_{a\cdots c}$ which is antisymmetric in all indices, of rank n consistent with the orientation of M, and satisfies

$$\epsilon_{a\cdots c}\epsilon^{a\cdots c} = \pm n! \tag{147}$$

(The plus sign must be used in (147) if the signature of g_{ab} contains an even number of $-$'s; otherwise the minus sign.) The tensor field $\epsilon_{a\cdots c}$ is called the *alternating tensor* associated with g_{ab} and the orientation of M. If the orientation of M is reversed, then the sign of $\epsilon_{a\cdots c}$ is reversed.

Let $\omega_{a\cdots c}$ be an $n-$form on M. Then

$$\omega_{a\cdots c} = \pm\frac{1}{n!}\,\epsilon_{a\cdots c}\,(\epsilon^{m\cdots n}\omega_{m\cdots n}) \tag{148}$$

Thus, the $n-$form is completely and uniquely determined by the value of the scalar field $\omega_{a\cdots c}\epsilon^{a\cdots c}$. In other words, on an oriented manifold with metric, there is a one-to-one correspondence between $n-$forms and scalar fields. We define the left side of

$$\int_U f\,dS = \int_U \frac{1}{n!}\,f\,\epsilon_{a\cdots c}\,dS^{a\cdots c} \tag{149}$$

by the right. We can integrate scalar fields over open regions of M. In particular,

$$\int_U 1\,dS = \int_U \frac{1}{n!}\,\epsilon_{a\cdots c}\,dS^{a\cdots c} \tag{150}$$

is called the *volume* of U. Thus, a metric gives rise, in particular, to the notion of volume.

Next, let $\mu_{a\cdots b}$ be an $(n-1)$–form on M.Then, setting

$$v^a = \epsilon^{ab\cdots c}\mu_{b\cdots c} \tag{151}$$

we have

$$\mu_{m\cdots c} = v^a \epsilon_{ab\cdots c} (\pm \frac{1}{(n-1)!}) \tag{152}$$

Thus, we obtain a natural, one-to-one correspondence between $(n-1)$–forms and vector fields on M. If S is an oriented sub manifold of M, of co-dimension one, we define the left side of

$$\int_S v^a \, dS_a = \frac{1}{(n-1)!} \int_S (v^a \epsilon_{ab\cdots c}) \, dS^{b\cdots c} \tag{153}$$

by the right side. In particular, if S is a metric submanifold we define the *area* of S by

$$\int_S v^a \, dS_a \tag{154}$$

where v^a is a unit normal vector.

We can now express n-form in terms of scalar fields, and $(n-1)$ forms in terms of vector fields. Hence, the exterior derivative (which takes an $(n-1)$–form to an n–form) should take a vector field to a scalar field. What is the expression for this action? Let $\omega_{a\cdots b}$ be an $n-1$ form, and v^a its corresponding vector field. Then, from the concomitant (54), we have

$$D(\omega)_{ma\cdots b}\, \epsilon^{ma\cdots b} = \nabla_m (\epsilon^{ma\cdots b}\, \omega_{a\cdots b}) = \nabla_m v^m \tag{155}$$

where ∇_m is the derivative operator defined by g_{ab}. Thus, the operation is simply that of taking the divergence using the derivative operator defined by g_{ab}.

Finally, substituting (155) into (143), we have

$$\int_U (\nabla_a v^a) \, dS = \int_{\partial U} v^a \, dS_a \tag{156}$$

This, of course, is the familiar form for Gauss' law.

Example 81. Let M be an oriented, three-dimensional manifold with metric. Then Theorem 22, for a 3–dimensional submanifold, is Gauss' law; for a 2–dimensional sub manifold is what is usually called Stokes' theorem; and for a 1–dimensional sub manifold is the line integral theorem (Example 78). Thus, the three standard integral theorems for a three-dimensional manifold are precisely the three versions of Stokes' theorem.

Exercise 214. Prove that $\nabla_a \epsilon_{b \cdots c} = 0$. where $\epsilon_{b \cdots c}$ is the alternative tensor, and ∇_a is the derivative operator defined by the metric. (Hint: Take a derivative of (147).)
Exercise 215. Find the volume of S^2 (with the usual metric).
Exercise 216. Find the area of S^2 consider as a submanifold of R^3 (with usual embedding and metric).
Exercise 217. Let M be an oriented n–dimensional manifold with metric. Extend the discussion above to obtain a correspondence between k–forms on M and $(n - k)$–forms. Express the action of the exterior derivative.

Appendix

Math 279 **Test** February 3, 1972
Due: February 19, 1972

The twelve problems below are arranged approximately is order of difficulty, from easiest to hardest. Do the hardest six you can.

1. Let ξ^a be a nonzero vector at a point p of the n−dimensional manifold M, and let $(\tilde{\xi}^1 \ldots, \tilde{\xi}^n)$ be n numbers, not all zero. Prove that there exists a chart with respect to which the components of ξ^a are $(\tilde{\xi}^1, \ldots, \tilde{\xi}^n)$.

2. Let $R_{abc}{}^d$ be a tensor at the point p of the paracompact manifold M, satisfying $R_{abc}{}^d = R_{[ab]c}{}^d$ and $R_{[abc]}{}^d = 0$. Prove that there exists a derivative operator on M whose Riemann tensor at p is this $R_{abc}{}^d$.

3. Find two distinct metrics on R^2 which define the same derivative operator. Find two distinct derivative operators on R^2 which have the same Riemann tensor.

4. Let η^a be a vector field on the manifold M such that $\mathscr{L}_\xi \eta^a = 0$ for every ξ^a. Prove that $eta^a = 0$

5. Prove Eqn. (31).

6. Let $\psi : M \to M'$ be smooth, and let U be an open subset of M'. Prove that $\psi^{-1}[U]$ is an open subset of M.

7. Let M and M' be manifolds. Find a natural diffeomorphism between $T(M \times M']$ and $TM \times TM'$.

8. Let M be a three-dimensional manifold with metric g_{ab}. Prove that, if the Ricci tensor vanishes, then so does the Riemann tensor.

9. Find a smooth, onto mapping $\psi : M \to M'$ such that $\vec{\psi}_p$ is an isomorphism for each point p of M, but such that ψ is not a diffeomorphism.

10. Find a connected manifold M such that, for any point p of M, $M - p$ is diffeomorphic with M.

11. develop an algebra of the concomitant (58) along the lines of Example 30.

12. For two totaly symmetric contravariant tensors $\alpha^{a_1 \ldots a_p}$ and $\beta^{a_1 \ldots a_q}$ at a point, set $\alpha \cap \beta = \alpha^{a_1 \ldots, a_p} \beta^{a_{p+1} \ldots a_{p+q}}$. A symmetric contravariant tensor $\gamma^{a \ldots, c}$ at this point will be said to be *prime* if it cannot be written as $\alpha \cap \beta$ with α and β

of nonzero rank. Prove that every symmetric contravariant tensor at the point can be decomposed into a product of primes, and that this decomposition is unique except for order and scalar factors.

Math 279 **Final Examination** February 29, 1972
Due: March 9, 1972

The twenty-four problems below are arranged into eight groups of three. The problems within each group are arranged in order of difficulty. Do the hardest ten you can, including at least one problem from each of the eight groups.

Group A.

1. Let μ_{ab} be an antisymmetric tensor at a point, and ν_a nonzero vector. Suppose $\mu_{[ab}\nu_{c]} = 0$. Prove that there exists a vector ω_b at the point such that $\mu_{ab} = \nu_{[a}\omega_{b]}$.

2. Let ϵ^{abcd} be the alternating tensor on an oriented four-dimensional manifold with positive-definite metric. Prove that $\epsilon^{abmn}\,\epsilon_{cdmn} = \pm\frac{1}{4}\delta^{[ac}\delta^{b]}{}_{d}$.

3. Let K^{abc} be a tensor at *a* point. Prove that K^{abc} can be written uniquely in the form $K^{abc} = \alpha^{abc} + \beta^{abc} + \gamma^{abc} + \delta^{abc}$ with $\alpha^{abc} = \alpha^{(abc)}, \beta^{abc} = \beta^{[abc]}, \gamma^{abc} = \gamma^{(ab)c}, \gamma^{(abc)} = 0$, $\delta^{abc} = \delta^{[ab]c}$, and $\delta^{[abc]} = 0$.

Group B.

4. Let p and q be distinct points of S^3. Prove that $S^{-}p - q$ is diffeomorphic with $S^2 \times R$.

5. Let M and M' be $n-$dimensional manifolds with vector fields ξ^a and ξ'^a, respectively, each of which vanishes nowhere. Let p be a point of M, and p' be a point of M'. Prove that there exists an open set U containing p, and U' containing p', such that there exists a diffeomorphism from U to U' which takes ξ^a to ξ'^a.

6. Find manifolds M and M' which are not diffeomorphic, such that $M \times R$ is diffeomorphic with $M' \times R$.

Group C.

7. Prove that the interior derivative of an $n-$form on an $n-$dimensional manifold always vanishes.

8. Let ξ^a and η^a be vector fields, with η^a vanishing nowhere. If $\mathcal{L}_\xi(\eta^a\eta^b) = 0$, prove that $\mathcal{L}_\xi\eta^a = 0$.

9. Let $\epsilon_{a\ldots c}$, be an $n-$form, vanishing nowhere, on an $n-$dimensional manifold. Let $\epsilon^{a\ldots c}$ be its inverse. Prove that $\mathcal{L}_\xi\epsilon_{a\ldots c} = 0$ if and only if $\mathcal{L}_\xi\epsilon^{a\ldots c} = 0$.

Group D.

10. On a manifold (of dimension greater than two) with metric, let $R_{abcd} = C\,g_{a[c}g_{d]b}$. Prove that $C = const$.

11. On a manifold with metric, a vector field ξ^a is said to define a *constant of the motion* if, for any geodesic with tangent η^a, $\xi_a\,\eta^a$ is constant

along the geodesic. Prove that ξ^a defines a constant of the motion if and only if ξ^a is a Killing vector.

12. Prove that, on a two-dimensional manifold with metric, there exists a scalar field α such that $R_{abcd} = \alpha g_{a[c} g_{d]b}$.

Group E.

13. Let $\psi : M \rightarrow M'$ be smooth. Prove that the subset of $M \times M'$ consisting of points of the form $(p, \psi(p))$ is a submanifold.

14. Let $\psi : M \rightarrow M'$ and $\Phi : N \rightarrow N'$ be smooth. Prove that $\Lambda : M \times N \rightarrow M' \times N'$ defined by $\Lambda(p, q) = (\psi(p), \Phi(q))$ is smooth.

15. Let $psi : M \rightarrow M'$ be smooth. When is ψ a one-to-one correspon-
$\leftarrow$
dence between covariant vector field on M' and those on M?

Group F.

16. Prove that the image of a cross-section of a vector bundle is a submanifold of the bundle space.

17. Prove that two derivative operators on a manifold are the same if and only if they define the same horizontal subspace of each tensor bundle.

18. Let M be a manifold with positive-definite metric. Find a natural positive-definite metric on the tangent bundle of M.

Group G.

19. Let ξ^a be a complete vector field on M, and p a point of M. Prove that ξ^a is complete on $M - p$ if and only if $\xi^a = 0$ at p.

20. Classify, the two-dimensional Lie algebras.

21. find two complete vector fields on a manifold whose sum is not complete.

Group H.

22. Let S (a subset of a manifold M) be a submanifold, and let T (a subset of S) be a submanifold of S. Prove that T is a submanifold of M.

23. Let S be the subset of R^3 consisting of points with $(x^1)^2 - (x^2)^2 - (x^3)^2 = 1$. Let g_{ab} be the metric on R^3 with components $g_{11} = -1$, $g_{22} = g_{33} = 1$, others zero. Find the extrinsic curvature and intrinsic metric of S.

24. Find a one-form ω_a on S^1 R such that $D(\omega)_{ab} = 0$, but such that, for some closed curve C, $\int_C \omega_a \, dS^a \neq 0$.

About the author

Robert Geroch is a theoretical physicist and professor at the University of Chicago. He obtained his Ph.D. degree from Princeton University in 1967 under the supervision of John Archibald Wheeler. His main research interests lie in mathematical physics and general relativity.

Geroch's approach to teaching theoretical physics masterfully intertwines the explanations of physical phenomena and the mathematical structures used for their description in such a way that both reinforce each other to facilitate the understanding of even the most abstract and subtle issues. He has been also investing great effort in teaching physics and mathematical physics to non-science students.

Robert Geroch with his dog Rusty

5768918R00089

Printed in Great Britain
by Amazon.co.uk, Ltd.,
Marston Gate.